Ergonomics of Carrying Loads by Women in Hills

About the Authors

Dr. Vijay Shree Dhyani is teaching personnel (Department of Family Resource Management,) College of Home Science G.B.P.U.A&T, Pantnagar, Uttarakhand. She has been involved in research as a scholar during her MSc and PhD research. She has also developed external frame backpack for carrying loads by women of Uttarakhand.

Dr. Promila Sharma is Professor and ex-head, Department of Family Resource Management, College of Home Science G.B.P.U.A&T, Pantnagar, Uttarakhand. She has been involved in research since 1983 and disseminated several technologies/techniques through research and extension work

Dr. T. C.Thakur is the ICAR National Professor, Department of Farm Machinery and Power Engineering. College of Technology, G. B. P. U. A. & T., Pantnagar, Uttarakhand. He was Commonwealth Scholar at the Cranfield Institute of Technology, England (U. K.) during 1981-85 to pursue his higher studies. He has over 38 years of experience in teaching, research and extension of various technologies. Dr. Thakur has published over 170 research papers/technical articles in National and International Journals. He has also published 16 research/extension bulletins and contributed to 18 book's chapters. He is the recipient of many honours and awards because of his outstanding contributions in the field of Agricultural Engineering. Dr. Thakur has developed several types of innovative hill implements and farm machinery for plain regions which are being patented and commercialized. He is on the role of many esteemed orgnizations in the capacity of member and advisor.

Ergonomics of Carrying Loads by Women in Hills

(For Drudgery reduction, Safety and Comfort)

Vijay Shree Dhyani
Promila Sharma
T.C. Thakur

CONCEPT PUBLISHING COMPANY PVT. LTD.
NEW DELHI-110059

ISBN-13 : 978-81-8069-930-6

First Published 2013

Published and Printed by

Concept Since 1974

Concept Publishing Company Pvt. Ltd.
A/15-16, Commercial Block, Mohan Garden
New Delhi-110059 (India)
T: 25351460, 25351794, *F*: 091-11-25357109
E: publishing@conceptpub.com, W: www.conceptpub.com
Editorial Office: H-13, Bali Nagar, New Delhi-110 015, India.

Cataloging in Publication Data--*Courtesy:* D.K. Agencies (P) Ltd. <docinfo@dkagencies.com>

Dhyani, Vijay Shree.
Ergonomics of carrying loads by women in hills : for drudgery reduction, safety and comfort / Vijay Shree Dhyani, Promila Sharma, T.C. Thakur.
p. cm.
Includes bibliographical references (p.) and index.
ISBN 9788180699306

1. Women employees—India—Garhwal (Region)—Economic conditions. 2. Women employees—India—Kumaun—Economic conditions. 3. Women employees Workload—India—Garhwal (Region) 4. Women employees—Workload—India Kumaun. I. Sharma, Promila, 1955-, joint author. II. Thakur, T. C., joint author. III. Title.

DDC 331.481330.95451 23

Dr. B. S. Bisht
Vice-chancellor

G.B.Pant University of Agriculture
and Technology, Pantnagar-263145
Distt.-U.S.Nagar
Uttarakhand
India

Foreword

Women comprise about half of the workforce and many of them are involved in agriculture in addition to their traditional domestic occupations. One of the reasons for higher women workforce in hilly areas is that most of the able bodied men migrate in plains for livelihood and leave not only task of running the household squarely on the shoulders of women but also most of the heavy agricultural tasks. Hill women spent most of their time carrying water, fuel wood, fodder materials from long distances which involve carrying loads on head, back, shoulders etc. These involve walking long distances in hilly regions with poor infrastructural facilities, poorly constructed roads with a lot of stones and pebbles, narrow pathways, full of twists and turns with slanting/steep slopes and scattered land holdings. All of these contribute to increase the physiological costs and physical loads to a great extent. It ultimately causes unbalancing of body with heavy loads due to which hill women suffer from various kinds of ailments such as backache, pain in shoulder and neck, and accidents and fall/slips causing serious injuries.

A unique feature of this book is that it takes into consideration the ergonomics of load carrying in a very systematic and concise manner so as to benefit the hill women and other load carrying population from injuries and related health hazards of load carrying. This really needs focus as hill women spend most of their time in load carrying activities.

I compliment Dr. Vijay Shree Dhyani, Dr. Promila Sharma and Dr. T.C.Thakur for their effort in publication of this important book. I am sure that the scientific information contained in the form of various chapters on different aspects of carrying loads and related aspects will be useful to planners, policy-makers, researchers and all those who are engaged in welfare and well-being of the load carrying population especially in hill areas and it will also help in designing tools and techniques of safe load carrying as they have already developed backpack taking into consideration the anthropometric measurement's of hill women and other ergonomic aspects..

Dr. B. S. Bisht
Vice-Chancellor

Preface

Most of the time hill women spend carrying water, fuel, wood, fodder materials from long distances which involve carrying loads on head, back, shoulders etc. Manual material handling activities such as lifting, pulling, pushing and carrying of weights have been recognized as being associated with a high prevalence of musculoskeletal disorders in the working population. Most extensive are the studies and recommendation or legislation on lifting conditions while knowledge regarding carrying loads is sparse. Besides the heavy work of carrying loads by hill women, the other major concern is that they vary tremendously in their anthropometric dimensions not only from men but also from other races. There is scanty information available on women in general and hill women in particular regarding load carrying studies, recommendation and anthropometric data. Scarcity of information on above aspects provided a solid reason to undertake women as the subject to investigate effects of modes, grades, loads and environmental conditions on physiological cost and perceived exertion as a research problem.

Considering the above facts and point of views, it was proposed to investigate changes in the physiological responses and perceived exertion due to loads, modes, grades and environmental conditions, most critical factors in the hill women's life. Generation of anthropometric data on hill women is required to develop appropriate technologies for their better health and well-being, and also tools and equipment either should be modified or designed keeping in mind the anthropometric dimensions and capabilities of hill women.

Diverse workforce in some occupations also make the anthropometric differences between occupations more pronounced, while new roles for women in the workforce make it likely that more occupation will show a broader range of body dimensions. Since different occupational groups perform different jobs, use of inappropriate anthropometric data for designing workplaces, systems and tools is a serious safety concern. This book ensures that human capabilities, limitations and awareness of the practical importance of physical size and body measurements are very crucial in designing tools and technologies. An effort was made to design ergonomic backpack for carrying loads in hill regions taking into consideration the ergonomic aspects.

Vijay Shree Dhyani
Promila Sharma
T.C.Thakur

Acknowledgements

The tenure of human in this world is supported by many others. Acknowledgements for a few might be just a trifle thing written on a piece of paper. But in the true sense it gives an opportunity to remember and express our token of thanks to people who in a way helped and supported to complete this record.

We wish to extend our deep sense of gratitude and sincere thanks to Dr. S.K. Tewrai, Professor, Department of Agricultural Economics, Dr. A.K. Shukla, Professor and Head, Department of Math and Statistics and Dr. S.K. Rastogi, Professor and Head, Department of Veterinary Physiology for their constant inspiration, valuable suggestions and sustained interest during the course of investigation and preparation of manuscript. We also express our sincere thanks to Dr. S.C. Tewari, for help during data analysis.

We extend our gratefulness to Dean, College of Post-graduate Studies, Dean College of Home Science for providing necessary facilities for conducting the research.

We are immensely grateful to Dean, Hill Campus Ranichouri of G.B.P.U.A. & T, Pantnagar for providing necessary help during the data collection and also to Punjab Agriculture University, Ludhiana for providing timely help for accessing library.

We duly acknowledge our sincere thanks to Doorbhash international, New Delhi for fabricating the prototype of external frame backpack.

We take this opportunity to extend our thanks to all the teaching and non-Teaching members of the Department of Family Resource Management for their help and cooperation.

It is difficult to express our feelings to our respective families for their constant support and encouragement

We cannot forget to appreciate the outstanding cooperation from friends and well-wishers.

Any omission in this brief acknowledgements does not imply ingratitude.

Pantnagar
July, 2012

Vijay Shree Dhyani
Promila Sharma
T.C.Thakur

Contents

List of Tables

List of Figures and Plates

Figures

Plates

Chapter -1

Introduction

Women comprise about half of the workforce and many of them are involved in agriculture in addition to their traditional household chores. In India, at present the women workforce in agriculture and allied sector is estimated to about 30 per cent of the total rural workforce of the country. The 1991 census recorded a high female workforce participation rate in some of the hill districts e.g. about 49.5 per cent of workforce in Almora district of Uttarakhand State *(Bose, 2000).*

One of the reasons for higher women workforce in hilly areas is that most of the able bodied men migrate in the plains for livelihood and leave not only task of running the household squarely on the shoulders of women but also most of the heavy agricultural tasks. The economy of hills is largely dependent on natural resources, and ever shrinking forest and water resources have hit women really hard. They spend longer hours gathering fuel, wood, fodder and fetching water. Then losses of forest cover have made agricultural activities more cumbersome. They often work for 14 hours a day and during peak season (September-October) when women's work day extend up to 15-16 hours a day.

Most of the time hill women spend carrying water, fuel, wood, fodder materials from long distances which involve carrying loads on head, back, shoulders, on lap of hands etc. These involve walking long distances in hilly regions with poor infrastructural facilities, poorly constructed roads with a lot of stones and pebbles, narrow pathways, full of twists and turns with slanting and steep slopes. All of these contribute to increase the physiological costs and physical loads to a great extent. It also causes unbalancing of body with heavy loads. Other debilitating factor involves scattered land holdings which again cause women to cover long distances in difficult environmental and geographical conditions, due to which hill women suffer from various kinds of ailments such as backache, pain in shoulder and neck, and fall or slips causing serious injuries.

Manual material handling activities such as lifting, pulling, pushing and carrying of weights have been recognized as being associated with a high prevalence of musculoskeletal disorders in the working population. Most extensive are the studies and recommendation or legislation on lifting conditions (NIOSH 1981, 1991, Christensen *et al.*, 1995) while knowledge regarding carrying loads is sparse.

It has been reported that the metabolic demand during load carrying increases linearly with the carrying weight (Soule and Goldman, 1969; Soule *et al.*, 1978; Keren *et al.*, 1981; Gordon *et al.*, 1983; Francis and Hobbler, 1986).

Musculoskeletal disorders are more common among women rather than men, especially in neck and shoulder disorders (Punnet and Herbert, 2000 and De Zwart *et al.*, 2001).

The physiological cost of carrying loads has been extensively investigated in India (Datta *et al.*, 1968; Datta and Ramanathan, 1970; Samantha and Chatterjee, 1981) and elsewhere (Goldman and Lampietro, 1962; Goldman 1965, Kanon and Belding, 1971). However, studies on permissible weight for carrying loads manually by men have been carried out by Malhotra *et al.* (1962), Malhotra and Sengupta (1965), Das and Saha (1966), Hughes and Goldman (1970) and Datta *et al.* (1975) but most studies did not consider the workers age, therefore no general statement about capabilities and limitations to carry loads is possible on the basis of available data (Samantha *et al.*, 1986).

The physical work is still present in many countries, especially in developing ones. The physiological analysis of the work load could be conducted through two different ways, referring to his/her energetic cost and/or to the mechanical power of the human motor. The purpose of ergonomics is to organize any physical professional task in order to maintain the workers as much as possible inside the limits of a fatiguing activity. The process of fatigue affects differently a 'general activity' and a 'local work.' If a worker is engaged in a general hard activity, like carrying load or pushing a trolley, an increase of oxygen consumption or/and an acceleration of the heart rate could be observed. This latter, allowing estimating the oxygen cost of the work is more affected if the work is performed in a hot environment. In order to establish the ratio between the data observed at the work place and the maximal working capacity of the organism, it is possible to experiment with the physiological effects of transitory increases of the work stress, by increasing the load or/ and the frequency of the repetition of the 'unit' of work done. When the task could be described in terms of local work, the correct methodology is based upon the general rules established respectively for the static and the dynamic work. In both the cases, it is possible to determine the critical rate of the muscular activity, the part of the muscular capacity engaged at the work place and the probable time for local exhaustion, in order to limit the duration of each period of work.

Human capabilities, limitations and awareness of the practical importance of physical size and body measurements is widely known. Diverse workforce in some occupations also make the anthropometric differences between occupations more pronounced, while new roles for women in the workforce make it likely that more occupation will show a broader range of body dimensions. Since different occupational groups perform different jobs, use of inappropriate anthropometric data for designing workplaces, systems and tools is a serious safety concern (Hsiao, 2002).

Besides the heavy work load of carrying loads by hill women, the other major concern is that they vary tremendously in their anthropometric dimensions not only from men but also from other races. There are very few anthropometric data (Nag Vargese, 1981, 1983; *et al.*, 1987; Vasta, 1996 and Oberoi, 1996) available on women (Gite and Singh, 1997) but anthropometric data on hill women (Uttarakhand) is negligible due to which mismatch occur between human dimensions and dimensions of tools and equipment that Uttarakhand women use. The major problem lies in the fact that most of the agricultural tools and equipment currently used by them were being designed according to the anthropometric measurements of men which causes various musculoskeletal disorders and injuries to them. As anthropometry is an integral part of human limitations and capabilities and the fact that the anthropometric dimensions vary significantly between men and women and also between the races, so it is essential to consider the gender and racial differences while designing the tools and equipment.

Variations in the anthropometric data exist due to ethnic diversity, population groups, geographical conditions, working conditions etc. For occupational injury prevention, anthropometric measurements are used to evaluate the interaction of workers with tasks, tools, machines, vehicles and personal protective equipment. Designs that are incompatible with anthropometric measurements of a workforce could result in undesired incidents. While literature has shown that the gender, age, race and geographical region are associated with difference in body dimensions (SAE International, 1998).

The anthropometric differences among races are greater than among nations. Different nations of the same race may also have varied body sizes and bodily proportions due to the differences in social and economic environment and it is evident through general observation that there are differences in anthropometric measurements between various regions and parts of the same nation. So it is imperative to find out whether there are significant differences in anthropometric measurements between Garhwal and Kumaon regions of Uttarakhand, and to compare the anthropometric dimensions of Uttarakhand women with anthropometric dimensions of other states.

Considering the above facts and point of views, it was proposed to investigate changes in the physiological responses and perceived exertion due to loads, modes, grades and environmental conditions, most critical factors in the hill women's life. Generation of anthropometric data on hill women is required to develop appropriate technologies for their better health and well-being, and also tools and equipment either should be modified or designed keeping in mind the anthropometric dimensions and capabilities of hill women. Therefore, the study was undertaken with the following specific objectives:

Objectives

1. To collect and analyse the anthropometric data of hill women of Uttarakhand.
2. To find out the health hazards, injuries and musculoskeletal disorders of hill women of Uttarakhand while carrying loads.
3. To find out the Heart Rate (HR), Oxygen Consumption Rate (OCR), Energy Expenditure Rate (EER), OCR/kg body weight, Metabolic Equivalents (METS), Rating of Perceived Exertion (RPE) and Body Part Discomfort Score (BPDS) of a few manual modes of carrying loads at different grades.
4. To find out the effect of different environmental conditions (temperature and humidity) on the HR and RPE of the least cost mode of carrying loads.
5. To suggest the work-rest allowance under different conditions.
6. To develop a gadget for carrying loads in hills and its effect on physiological responses and RPE of women subjects.

Hypotheses

1. The mean anthropometric measurements of Garhwal and Kumaon region are equal.
2. The effect of various modes of carrying loads, load and grade on HR, increase in HR, OCR, EER, OCR/kg body weight, METS and RPE is same.
3. There exists a relationship between (HR and OCR), (HR and RPE) and (OCR and RPE).
4. The HR and OCR are a function of mode, load and grade.

Limitation

1. The study was limited to women only.

Chapter-2

Review of Literature

Any working system in ergonomics consists of three components *viz.* the man, machine and the working environment. The present study is concerned with the ergonomic studies on various modes of carrying loads on uphill and dawn hill. A brief review of the studies relevant to the topic has been presented in the following sections :

1. Selection of subjects.
   (i) Age of subjects
   (ii) Weight of subjects
2. Anthropometric measurements
3. Physiological responses of load carrying
4. Effect of environment on physiological responses
5. Health hazards and injuries
6. Effect of backpacks on physiological responses
7. Salient findings from review
8. Conclusion

1. Selection of Subjects

The subjects have an important role in ergonomic studies, hence they should be medically fit. The subjects are selected on the basis of a detailed anamnesis, including the physical fitness, age, weight and should be real representative of the population as well as the number of the subjects that can be considered as valid for the study.

(i) Age of subjects

Grandjean (1982) stated age dependent changes in muscular force developed and that the peak of muscle strength for both men and women is reached between the age of 15 to 35 years. The workers aged between 50 and 60 years can exert only 75 to 85 per cent of muscular strength. Rodahl (1989) quoted that the maximum heart rate declines with age.

Mc Ardle et al. (1994) reported that the body functions generally improve rapidly during childhood and reach to maximum between 20 and 30 years of age. Thereafter, a gradual decline in the functional capacity occurred with advancing years.

Gite and Singh (1997) found that the maximum strength can be expected from the age group of 25 to 35 years. Later on Nigg and Herzog (1999) also reported that the maximum muscle strength and cross sectional area of muscle are greatest for the age group of 25 to 35 years.

(ii) Weight of subjects

Morrison and Harrington (1962) reported that the ischial tuberosities (lower most projections of the pelvis) normally carry the weight of the upper part of the body and these projections spaced only 108 to 114 mm apart in the human adult and are capable of transmitting large force directly and uniformly into spinal column.

2. Anthropometric Measurement

Ducharme (1973) had also shown that the average women is about the same height as the shortest five per cent of men and that the average man is about the same height as the top five per cent of women.

Chand and Singh (1978) reported the mean values and standard deviations of body fat and lean body mass of 65 young Indian women aged 20–25 years as 9.00 kg (s.d. 6.14) and 36.02 kg (s.d. 3.99), respectively. Body fat per cent was 18.72 (s.d. 10.48). On selective stepwise multiple linear regression analysis, the measurements selected for predicting body fat percentage were maximum at thigh girth, chest diameter, height up to gluteal furrow and shoulder diameter, all four together give a multiple correlation coefficient of 0.88. For predicting lean body mass, the measurements selccted were stature, upper arm length, lower thigh girth and chest diameter, all four together give a multiple correlation coefficient of 0.84.

Gupta et al. (1983) carried out anthropometric measurement of forty subjects of the Punjab. They measured standing height, seat height, seat depth, hip breadth, and elbow rest height and body weight from seat design point of view. They reported that values of standing height of the 5th, 50th and 95th percentile were 155.20, 168.50, 179.00 cm, respectively. They also found a linear relationship between standing height and other dimensions except hip breadth. It was observed that the data of Indian workers were significantly different from the data of U.S and Dutch people; therefore it could not be applied in India.

Gite and Yadav (1989) reported that in Indian agriculture, hand tools, animal-drawn equipment and tractor/power operated machinery are extensively used for various operations. The equipment are either operated or controlled by human workers. Use of anthropometric data can help in the proper design of equipment for better efficiency and more human comfort. They identified 52 body dimensions for the design of equipment and proposed that extensive surveys should be carried out in different regions of the country to generate the necessary data useful in farm machinery design.

Pheasant (1996) reported that the high prevalence of musculoskeletal disorders in women compared to the men may also depend on the design of workplaces and hand tools, which are often based on anthropometric data of men.

Hsiao (2002) studied anthropometric differences among occupational groups and concluded that the body segment measurement of some occupational groups differed significantly. Occupational factors seem to play an important role on the differences, in addition to age, gender and ethnicity effects.

Pund et al. (2002) carried out an extensive survey to collect anthropometric data of male and female agricultural workers involved in different agricultural operations in Saurashtra region (western India). They found that average stature of male and female workers recorded were 163.1±4.92 and 150±5.60 cm, respectively.

Nag *et al.* (2003) conducted a survey of 95 women working in informal industries (beedi, agarbatti and garment making) were selected and fifty one hand measurements of the right hand (lengths, breadths, circumferences, depths, spreads and clearances of hand and fingers) were taken. They found that hand breadths, circumferences and depths were approximately normally distributed, with some deviation in case of the finger lengths. Hand length was significantly correlated with the fist, wrist and finger circumferences. The fist and wrist circumferences, in combination, were better predictors of hand length. The hand lengths, breadths and depths, including finger joints of the Indian women studied were smaller than those of American, British and West Indian women. The hand circumferences of the Indian women were also smaller than the American women. Grip strengths of Indian women (20.36+/-3.24 kg) were less than those of American, British and West Indian women.

Lin et al. (2004) studied ethnic differences in thirty-three anthropometric characteristics among four races i.e. Chinese, Japanese, Korean, Taiwanese, in East Asia and reported that there is a significant morphological difference among these people in the same region. The Chinese body shape has a narrower body with mid-range limbs., whereas Japanese body shape is wider with shorter limbs. The Korean body shape is mid-range among the four, but the upper limbs are longer and Taiwanese body shape has wide shoulder and narrow hip with large hands and long legs.

Panchan *et al.* (2004) proposed two simple and a dimensional indices to use in the quality control of anthropometric surveys. Such indices are applicable only to a few anthropometric variables but the results found by this control can be an indication of the quality of the complete measurement process. The indices are defined as [(se=100x (shoulder height-elbow height)-arm length/ (arm length), and *SES* index is the same when the measures are obtained in sitting position. The observer must take those measurements again when the index values are higher than a selected threshold value before approving or rejecting the measure.

Dewangan (2005) conducted a survey of thirty-three anthropometric dimensions from 280 male farmers belonging to 7 states of the eastern region. Range, mean, standard deviation, coefficient of variation, 5th, 50th and 95th percentile for each anthropometric dimension were calculated for the entire region. The anthropometric dimensions of farm youths of the north eastern region were compared with those of the northern, central, eastern, southern and western regions of India. It was found that the farm youth of the north eastern region had most of the body dimensions lower than those from other regions except southern and eastern regions of India. The anthropometric dimensions of youth of the north eastern region of India were also compared with those of China, Japan. Taiwan, Korea, Germany, Britain and USA. It was found that all the dimensions were lower than those from other parts of the world and concluded that for efficient design of improved agricultural machines for higher food production, the anthropometric data of the operators are very essential.

Shah *et al.* (2006) compared the Ideal Body Weight (IBW) formulae and published height-weight tables for healthy adults in the United States of America with the Body Mass Index (BMI) of 22 kg/m^2, which is associated with lowest mortality. Numerous formulae have been evolved to determine IBW, but there are no published studies comparing the validity of formulae with each other or comparing formulae with BMIs. Data from height-weight tables, weight for specific heights determined by IBW formulae, and weight for BMIs of 20, 22, 25, and 30 kg/m^2 at different heights were plotted for both men and women. Slopes based on a range of heights were determined for each formula using relational database software. The value for each slope obtained by linear regression was compared with the BMIs to determine which fit is best with BMI of 22 kg/m^2. Most height-weight tables and formulae predicted IBWs within the BMI range of 20–25 kg/m^2. However, for shorter heights the formulae were closer to BMI of 20 kg/m^2 and for taller heights, were closer to BMI of 25 kg/m^2. Height-weight tables, slopes were closer to the BMI slopes than formula slopes. Robinson's formula appeared to be the best equation for calculating desirable/ healthy weights in men; however, no formulae predicted close to a BMI of 22 kg/m^2 for women. Thus, in practice it might be more useful to use BMI ranges instead of IBW formulas for men and women.

Adak (2006) collected data on nutritional status, body weight, height, and sitting height from 11,496 adult males, aged between 18 and 62 years, belonging to 38 different populations of five major social groups (scheduled tribes, scheduled castes, other backward castes, general castes, and Muslims) of Central India. They found that the level of malnutrition is the lowest among the general castes. The opposite is the case with the scheduled castes and scheduled tribes. There was variation in weight and BMI but that there was no marked variation in other anthropometric variables. The five social groups differed more in size distance than in shape distance.

3. Physiological Responses of Load Carrying

When any work or activity is performed it manifest into physical exertion to the body and is characterized by high energy consumption and stress on the heart and lungs. The physiological response, in any work or operation is expressed in terms of cardio-respiratory response. The parameters measured are the heart rate, oxygen consumption rate and pulmonary ventilation rate. The heart rate indicates the total stress on the body.

Christensen (1953) classified the physical work on the basis of heart rate, oxygen consumption and energy expenditure rate for men and women as shown in Table 2.1:

Singh and Kaul (1972) investigated six modes of load transport and reported that the pulling load with face in direction of travelling was most demanding physiologically requiring oxygen consumption of 2.14 l/min whereas load carried on back supported by hands over shoulders and that carried on head demanded 1.42 and 1.41 l/min oxygen consumption and further reported that lowest value of 1.25 l/min was found during carrying load on back supported by hands from below.

Table 2.1 : Classification of physical work by Christensen (1953)

S. No.	*Classification of physical work*	*Heart rate (beats/min)*	*OCR (l/min)*	*Energy expenditure rate (kcal/min)*	
			Men		*Women*
1.	Unduly Heavy	151-175	>2.5	>12.5	>9.5
2.	Very Heavy	126-150	2.0-2.5	10.0-12.5	7.5-9.5
3.	Heavy	101-125	1.5-2.0	7.5-10.0	5.5-7.5
4.	Moderate	76-100	1.0-1.5	5.0-7.5	3.5-5.5
5.	Light	75	0.5-1.0	2.5-5.0	1.5-3.5
6.	Very Light	75	<0.5	<2.5	

Borghols *et al.* (1978) reported that dynamic exercise increases the oxygen uptake, heart rate and pulmonary ventilation linearly with the amount of weight carried. The increase being independent of the dynamic work load. In this study it was found that each kilogram extra weight increased oxygen uptake by 33.5 ml/min, heart rate by 1.1 beats/min and pulmonary ventilation by 0.6 l/min. In dynamic work for loads higher than 50 per cent of VO_2 max the relationship between weight carried and pulmonary ventilation is no longer a linear one. For the subjects studied it was assumed that oxygen consumption per kilogram weight was not different for body weight or extra weight carried.

Pimental and Pandolf (1979) developed a formula for prediction of energy consumption based on variables like body mass, walking velocity and grade. However; such calculations cannot differentiate between different modes of carrying load, which has been reported to be of great significance (Balogun, 1986; Datta and Ramanathan, 1971).

Myles and Saunders (1979) conducted a study on nine subjects walked on a treadmill with load (weights) equal to 10 per cent and 40 per cent of body weight carried on the back. Although the speed of the treadmill was selected so that the measured oxygen consumption (VO_2) was the same for both load conditions. The heavier load placed an extra strain on the cardiopulmonary system and was perceived by all subjects as harder work than the lighter load. When the subjects worked at their own pace, walking on a level road or climbing stairs with load (weights) equal to 10 per cent and 40 per cent of body weight, they compensated for the heavier load by decreasing walking speed or climbing rate. Although the energy costs calculated from walking speed, body and load (weights) for self-paced walking and the external work of stair climbing were the same for both load conditions, the heavier load was again perceived as harder work.

Robertson *et al.* (1982) studied differentiated local ratings of perceived exertion from the legs and central ratings from the chest and oxygen consumption were determined during load carriage in seven young women. Subjects walked for 6 min at 3.22, 4.83, 6.44, or 8.05 km h^{-1} carrying (i) no load, (ii) a load equal to 7.5 per cent of body weight (mean: 4.66 kg) and (iii) a load equal to 15 per cent of body weight (mean: 9.32 kg). Thus, each subject underwent 12 separate tests. The external loads were in the form of lead pellets carried in a plastic scuba belt worn around the waist. A differentiation threshold was found at 6.44 km h^{-1} for the 0 per cent and 7.5 per cent loads and at 4.83 km h^{-1} for the 15 per cent load. At speeds below the threshold, the perception of exertion was similar in the legs, chest and overall. At higher speeds, exertion was perceived to be more intense in the legs than overall and less intense in the chest than overall, suggesting that the local legs signal was the dominant factor in shaping the overall sensation of exertion. The oxygen uptake was greater for 15 per cent load than for either 0 per cent or 7.5 per cent loads, but was similar for 0 per cent and 7.5 per cent loads. Findings suggested a critical weight limit for external loads that could be transported without increasing the metabolic cost beyond that required to move the body weight alone. This limit fell between 7.5 per cent and 15 per cent of the body weight. When oxygen uptake was expressed per kg of total weight transported, there was no loss of metabolic efficiency while carrying loads up to 15 per cent of the body weight.

Balogun (1986) examined the cardio-respiratory responses when carrying external loads on the head and by yoke. Ten physically fit male students carried three different external loads (11.6, 16.1 and 20.6 kg) on a level motor-driven treadmill at varying speeds (0.89, 1.23 and 1.57 ms^{-1}), using three different modes of load carriage systems, i.e., head pack (HP), transverse yoke (TY), and frontal yoke (FY). Physiological measurements were obtained for heart rate (HR), pulmonary ventilation (PV), oxygen consumption (VO_2), ventilation equivalent (VE) and oxygen pulse (OP). The analysis of variance showed significant main effects (mode, load and speed) and three interaction effects in the HR, VE and VO_2 responses. Only the speed main effect was significant ($P < 0.05$) for VE. For the OP variable, significant F ratios were found for the mode ($P < 0.05$), load and speed ($P < 0.01$) main effects, but not for the interaction between mode, load and speed ($P < 0.05$). Scheffe's simple cell mean contrasts indicated that the cardio respiratory (HR, VE and VO_2) responses for the TY and HP modes were lower ($P < 0.05$) than the FY while transporting heavy (16.1 and 20.6 kg) loads at fast (1.23 and 1.57 ms^{-1}) walking speeds. Irrespective of the walking speed, the cardio respiratory responses while transporting loads with the TY and HP modes were not statistically different ($P < 0.05$). The findings suggest that load transportation by the frontal yoke is ergonomically inadvisable.

Maloiy et al. (1986) found that both the African Luo and Kikuyu women could carry loads of up to 20 per cent of their body weight without increasing their rate of energy consumption. For heavier loads there was a proportional increase in energy consumption, i.e. a 30 per cent load increased energy consumption by 10 per cent, a 40 per cent load by 20 per cent and so on. They suggested that some element of training and/or anatomical change since childhood might have allowed these women to carry heavy loads economically.

Charteris *et al.* (1989) and Maloiy *et al.* (1986) reported that the metabolic cost while walking with loads did not necessarily increase if the load was less than 20 per cent of the subject's body mass and termed this phenomenon as 'free-ride' and referred to 'free-ride' as carrying loads on the head.

Gleim and Nicholas (1989) compared oxygen consumption (VO_2) of treadmill walking/jogging in water to a dry treadmill at speeds of 40.23 to 160.9 m/min in 13.4 m/min increments. At speeds greater than or equal to 53.6 m/min, ankle depth, below knee, mid thigh, and waist depth walking/jogging in water significantly elevated VO_2 and heart rate (HR) above dry treadmill walking ($P < 0.05$). At speeds greater than or equal to 134.1 m/min, VO_2 of waist depth jogging was not significantly greater than dry jogging. These findings showed no gender specificity. Treadmill walking/jogging in waist depth water at temperatures of 30.5°C and 36.1°C was compared to dry treadmill walking in five subjects. The rate of increase of HR compared to VO_2 was significantly greater at 30.5°C than dry walking, and greater at 36.1°C than 30.5°C ($P < 0.05$). Treadmill walking in water can double the oxygen cost of movement depending on the depth and speed, and the response to increasing speed is nonlinear. Water temperature affected the relationship of HR to VO_2 at waist depth, suggesting that water temperature can add a significant thermal load to the cardiovascular system. Metabolic and cardiovascular demands of treadmill walking/jogging in water must be considered when using this modality since greater external work results at much lower speeds than on land.

Pivarnic and Sherman (1990) reported significantly lower heart rate responses at grades of -5 and -10 per cent than at +5 and +10 per cent in aerobically trained young men and women during both walking and slow jogging.

Patton *et al.* (1991) determined the energy cost of prolonged treadmill walking (fixed distance of 12 km) at speeds of 1.10 ms^{-1}, 1.35 ms^{-1}, and 1.60 ms^{-1}, unloaded (clothing mass 5.2 kg) and with external loads of 31.5 and 49.4 kg. Fifteen male subjects performed nine trials in random order over a 6-week period. Oxygen uptake (VO_2) was determined at the end of the first 10 min and every 20 min thereafter. A 10-min rest period was allowed between each 50 min of walking. No changes occurred in VO_2 over time in the unloaded condition at any speed. The 31.5 and 49.4 kg loads, however, produced significant increases (ranging from 10 to 18%) at the two fastest and at all three speeds, respectively, even at initial exercise intensities less than 30 per cent VO_2max. In addition, the 49.4 kg load elicited a significantly higher ($P<0.05$) VO_2 than did the 31.5 kg load at all speeds. In trials where VO_2 increased significantly over time, predicted values underestimated the actual metabolic cost during the final minute by 10–16 per cent. They concluded that energy cost during prolonged load carriage is not constant but increases significantly over time even at low relative exercise intensities. It was further concluded that applying the prediction model which estimates energy expenditure from short-term load carriage efforts to prolonged load carriage can result in significant underestimations of the actual energy cost.

Wanta *et al.* (1993) examined the heart rate and oxygen uptake (VO_2) responses of ten fit young men at two different speeds (90 and 105 m/min) on a level grade as well as downhill grades that changed in 3 per cent increments to 18 per cent and reported that responses were curvilinear with the lowest value for heart rate occurring between -6 per cent and -9 per cent grade and between -9 per cent and -12 per cent for VO_2.

Heglund *et al.* (1995) reported the lowest values of physiological cost for carrying head-supported loads as the Sherpa of the Himalayas or East African women.

Willems *et al.* (1995) showed that the energy cost of walking mainly depend on alteration of internal work done by locomotive muscles, and such internal work done by those muscles linearly increased with an increase in walking speed.

Knapik *et al.* (1996) reviewed the biomedical aspects of transporting loads in packs and offered suggestions for improving load-carriage capability. Locating the load mass as close as possible to the body centre of gravity appeared to result in the lowest energy cost when carrying a pack. Thus, the double pack (half the load on the front of the body and half the load on the back) had a lower energy cost than the backpack alone. However, backpacks alone provided greater versatility in most situations. The energy cost of walking with backpack loads increased progressively with increases in load mass, body mass, walking speed or grade and type of terrain. Training with loads resulted in greater energy efficiency since walking with backpack loads over several weeks decreases energy cost. Load-carriage speed could be increased with physical training that involves regular running and resistance training.

Robergs *et al.* (1997) measured heart rate in a group of recreational runners while running on grades of 0, -1.8, -3.6 and -5.4 per cent at three self-selected speeds, although not analyzed statistically, heart rate at any given speed became lower with progressively steeper grades.

Laursen *et al.* (2000) studied a biomechanical model for prediction of whole body energy turnover from kinematic and anthropometric data during load carrying. Total loads of 0, 10 and 20 kg were carried symmetrically or asymmetrically in the hands, while walking on a treadmill (4.5 km h^{-1}) horizontally, uphill or downhill at slopes of 8 per cent. Mean values for the directly measured oxygen uptake ranged for all the trials from 0.5 to 2.1 liter min^{-1}), and analysis of variance showed significant differences in respect of slope, load carried and symmetry. Heart rate was lowest during unloaded downhill walking (mean: 73 beats/min, range: 64-80 beats/min) and highest when carrying 20 kg asymmetrically uphill (mean: 130 beats/min, range: 113-152 beats/min). The heart rate data correlated closely with oxygen uptake (r^2 =0.778) and showed the same significant differences between trial conditions as did oxygen uptake. The close relationship between energy turnover rate measured directly and estimated based on biomechanical model justified the assessment of the metabolic load from kinematic data. The calculated values of oxygen uptake based on biomechanical model correlated significantly with the directly measured values, fitting to the line Y=0.990 X+0.144, where Y is the estimated oxygen uptake and X is the measured oxygen uptake in liters per minute.

Knight and Caldwell (2000) studied pole and no-pole conditions during uphill backpacking, which was simulated on an inclined treadmill with a moderately heavy (22.4 kg, 30 per cent body mass) backpack. Physiological measurements of oxygen consumption, heart rate, and RPE were taken during 1 h of backpacking in each condition, along with joint kinematic and electromyographic comparisons from data collected during a third test session. The results showed that although imposing no metabolic consequence, pole use elicited a longer stride length (1.27 vs 1.19 m), kinematics that were more similar to those of unloaded walking, and reduced activity in several lower extremity muscles. Although pole use evoked a greater heart rate (113.5 vs 107 beats/min), subjects were backpacking more comfortably as indicated by their ratings of perceived exertion (10.8 vs 11.6). The increased cardiovascular demand was likely to support the greater muscular activity in the upper extremity, as was observed in triceps brachii.

By redistributing some of the backpack effort, pole use alleviated some stress from the lower extremities and allowed a partial reversal of typical load-bearing strategies.

Falola *et al.* (2000) compared optimization characteristics during loaded or unloaded walking. Energy cost and stride characteristics were measured for 10 subjects with and without a load on the trunk of the body at 10 per cent of the body weight during four sessions. The first two sessions represented free walking at 2.5, 3, 3.5, 4, 4.5, and 5 km/h. The last sessions represented free vs forced walking at constant speed (preferred frequency and ±10 preferred frequency and ±20 per cent of preferred frequency). Their results showed an effect of load on energy cost of walking but no effect on the optimal speed for stability. Furthermore, when carrying a load, the subject did not adopt systematically the speeds that minimize physiological cost. The findings suggested the necessity to consider this effect to prevent gait disturbance and maintain the health benefits of walking.

James *et al.* (2004) studied cardiovascular and metabolic responses in healthy older and younger individuals during downhill walking and reported that older subjects had significantly greater heart rate, systolic B.P., diastolic B.P. and rating of perceived exertion during uphill, level and downhill walking compared to younger subjects. They observed a curvilinear response for heart rate, systolic B.P. diastolic B.P., and VO_2 with progressively steeper negative grades. These measures were highest at 5 per cent grade and lowest between -5 and -19 per cent grade.

Hooper *et al.* (2004) reported that percent maximum heart rate (HR max) and per cent maximum oxygen consumption (VO_2 max) increased incrementally for treadmill grades of 5 per cent to 7.5 per cent to 10 per cent for both forward walking and backward walking ($P < 0.01$). For each of the 3 treadmill grades, per cent HR max and per cent VO_2 max were 17 per cent to 20 per cent higher for backward walking than for forward walking ($P < 0.01$). No statistically significant interactions were detected between direction of walking and treadmill grade. Backward walking on a treadmill at 67.0 m/min (2.5 mph) and grades of 5 per cent, 7.5 per cent, and 10 per cent elicits a greater per cent HR max and per cent VO_2 max than does forward walking under the same conditions.

Abe *et al.* (2004) examined the effects of load, load position, and walking speed on the energy cost of walking per unit distance (Cw: ml/kg/m). Eight young male subjects walked on a treadmill at various speeds with and without load in the hands, on the back and on the legs. The Cw values were determined from the ratio of 2-min steady-state oxygen consumption (VO_2) above resting value (net VO_2) to the walking speed (v): Cw = net VO_2/v. They concluded that an energy-saving phenomenon would be observed due to an interaction between rotative torque around the centre of body mass and excessive burden on the lower muscles as a function of speed.

Bastien *et al.* (2005) studied the effect of speed and load on the energy expenditure of walking. The O_2 consumption and CO_2 production were measured in ten subjects while standing or walking at different speeds from 0.5 to 1.7 m s^{-1} with loads from 0 to 75 per cent of their body mass (M_b). The loads were carried in typical trekker's backpacks with hip support. Their result showed that the mass-specific gross metabolic power increased curvilinearly with speed and was directly proportional to the load at any speed. For all loading conditions, the gross metabolic energy cost (J kg^{-1} m^{-1}) presented a U-shaped curve with a minimum at around 1.3 ms^{-1}. At that optimal speed, a load up to 1/4 M_b seemed to be appropriate for long-distance walks. In addition, the optimal speed for net cost minimization is around 1.06 ms^{-1} and was independent of load.

Chung *et al.* (2005) investigated the physiological workload through heart rate and oxygen uptake involved in backpack-mode carrying, especially focusing on the effects of load (weight) and stairway transport. The results showed that stairways involved an increased physiological burden and that a load of 60 kg entailed a significantly higher physiological cost than carrying a load of 40 kg. They concluded that although backpack-mode carrying has some biomechanical advantages, the worker should carry a load of less than 40 kg to avoid a high physiological load. They also recommended that during backpack-mode carrying, a delivery person should make more trips with a light load rather than fewer trips with a heavier load.

4. Effect of Environment on Subject

Suggs and Splinter (1961) investigated effect of environment i.e. temperature, relative humidity and radiation on work load. It was reported that a subject was exposed to a mean radiant temperature of 70°C experienced a heart rate increase of 8-9 beats/minute above base values taken at mean radiant temperature of 31°C. On the basis of subject's normal response to bicycle ergometer work, this increase was equal to workload of 53.7kg-m/min. It was also reported that at a low relative humidity (about 30 per cent) the allowable workload was not affected by relative humidity. The temperature above this level increased the allowable workload. Pulse rate and work out-put linearity was not disturbed under constant environmental conditions.

Pirnary *et al.* (1969) studied the relationship between the heart rate and body temperature during muscular exercise in hot environment. It was found that on an average, the heart rate increased by 1.0 beat/min for each degree rise beyond 32.2 °C of body temperature.

Ishida (1975) reported results by estimating the physiological response (heart rate, respiration rate, rectal temperature and RMR i.e. relative metabolic rate to different degrees of walking conditions during the 10-minute exercise. The work intensity (RMR) of uphill walking with a certain load under hot environment can be changed by varying the walking speed and the load according to the slope gradient. As there existed a gap between RMR 5 and 6 in parallel with the other physiological responses, it is preferable to keep the walking condition under RMR 6, as viewed from the change of the other walking conditions. The heart rate as an indicator of the work load must be kept below 130 beats/min for a safe uphill walking of under RMR 6.

Gupta *et al.* (1977) studied the effect on metabolic responses during sub-maximal and maximal work load in dry and humid heat. It was reported that there was no significant increase in the heart rate with increase in temperature (29-40 °C) and relative humidity (50-55 per cent) for fixed work load of 600 kg-m/min. But maximum oxygen consumption (VO_2 max) showed a significant fall in very hot and extremely hot conditions with greater fall in humid conditions than in dry environment.

Thakur and Das (1978) investigated the effect of environment and modes of operation on rate and limitations of manual work output inside and outside the laboratory. Leg and hand cranking modes of operation were selected on the bicycle ergometer, and PVR, heart rate and oxygen consumption rate were measured at various work loads on the ergometer. It was found that all the parameters increased linearly with workload on the ergometer. Hand cranking was found more demanding physiologically than leg cranking. It was also found that subjects felt uncomfortable and showed early sign of fatigue in hot conditions. It was concluded that heart rate was more reliable physiological parameter for measurement of stress in environmental conditions and also for comparing the subjects at different modes of operation.

Germain and Edward (1985) reported that environmental conditions have a considerable impact on human performance and productivity in the construction industry. Yet, very little has been written on the relationship between environmental factors and productivity related to construction.

Eschenbacher *et al.* (1992) evaluated the potentially confounding effects of differences in temperature and humidity. They exposed 8 normal male subjects and 8 male subjects with asthma before and during 6 h exposure periods on 5 separate occasions i.e. cold, dry (10°C, 10 per cent relative humidity); cold, humid (10 °C, 50 per cent relative humidity); normal ambient (22°C, 40 per cent relative humidity); hot, dry (37°C, 15 per cent relative humidity); and hot, humid (37°C, 60 per cent relative humidity). The exposure period included a 12 min exercise on a cycle ergometer. No significant change was found in spirometry, airways resistance or diffusing capacity for either group of subjects at rest alone over the 6 h period of exposure for any exposure condition. They also found that the work performance (expressed as the external work performed divided by the oxygen consumed) decreased for the subjects in both groups at the conditions of higher temperature (37°C) compared with the lower temperature (10°C). These results confirmed that controlling the conditions of temperature and humidity is essential in chamber studies, field studies, or epidemiologic evaluations.

5. Related Health Hazards and Injuries

Knapik *et al.* (1996) reported that common injuries associated with prolonged load carriage include foot blisters, stress fractures, back strains, metatarsalgia (foot pain), rucksack palsy (shoulder traction injury) and knee pain. Closed-cell neoprene insoles and use of an acrylic or nylon sock, combined with a wool sock, reduce blister incidence. A framed pack with a hip belt reduced the incidence of rucksack palsy. Backpack load carriage can be facilitated by lightening loads, optimizing equipment, improving load distribution and by preventive action aimed at reducing the incidence of injury. Compared to backpacks, double packs produced fewer deviations from normal walking.

Emily (2003) reported that spinal pain in young people is a significant source of morbidity in industrialized countries. The carriage of posterior loads by young people has been linked with spinal pain, and the amount of postural change produced by load carriage has been used as a measure of the potential to cause tissue damage. The purpose of this review was to identify, appraise and collate the research evidence regarding load-carriage related postural changes in young people. A systematic literature review sought published literature on the postural effects of load carriage in young people. Sixteen databases were searched, which covered the domains of allied health, childcare, engineering, health, health-research, health-science, medicine and medical sciences. Two independent reviewers graded the papers according to Lloyd-Smith's hierarchy of evidence scale. Papers graded between 1a (meta-analysis of randomized controlled trials) and 2b (well-designed quasi-experimental study) were eligible for inclusion in this review. These papers were quality appraised using a modified Crombie tool. The results informed the collation of research evidence from the papers sourced. Seven papers were identified for inclusion in this review. Methodological differences limited our ability to collate evidence. Evidence based recommendations for load carriage in young people could not be made based on the results of this systematic review, therefore, constraining the use of published literature to inform good load carriage practice for young people.

6. Effect of Backpacks on Subjects

Legg and Mahanty (1985) also used two other subjective perceptual methods to try and distinguish between modes of load carriage. These were (i) a 100 mm Visual Analogue Scale (VAS) of perceived discomfort, and (ii) a modified Corlett and Bishop (1976) regional discomfort scale. The latter was a score determined as the sum of separate scores for all the areas of discomfort reported by each participant using a VAS for each of the twelve body regions. These perceptual subjective methods also failed to distinguish significantly between backpacks with or without a frame.

Lloyd and Cooke (2000) studied the energy expenditure associated with load carriage using a traditional rucksack and a new rucksack design, the AARN rucksack, which incorporates front balance pockets. Nine volunteers walked at 3 km/h at various uphill and downhill gradients on a treadmill without a load and with a load of 25.6 kg in each of the rucksacks. The oxygen consumption associated with both of the loading conditions was significantly ($P < 0.001$) higher than that associated with unloaded walking at all downhill gradients tested, although there was no significant difference between the two loading conditions. During the uphill gradients the oxygen consumption associated with the AARN pack was significantly ($P < 0.05$) lower than that associated with the traditional pack at the 0 per cent, 5 per cent, 10 per cent and 20 per cent gradients. The mean (per cent) differences at these gradients, expressed in ml/kg/min were 1.18 (9 per cent), 1.45 (8 per cent), 1.76 (8 per cent) and 1.88 (6 per cent), respectively. On an average for the whole protocol, the oxygen consumption associated with the AARN rucksack was 5 per cent lower than that associated with the traditional rucksack as 17.28 (s.d:7.46) ml/kg/min for the AARN pack and 18.20 (s.d:7.84) ml/kg/min for the traditional pack. The findings of the study suggested that a load carriage system that allows the load to be distributed between the back and front of the trunk is more appropriate for carrying relatively heavy loads than a system that loads the back only.

Stephen (2001) reported that numerous studies have used subjective perceptual methods to provide information about differences in backpack design. This study compared a Single-Strap Backpack (SSB). with a conventional Double-Strap Backpack (DSB) using subjective perceptual methods which included category ratio scale ratings of perceived regional comfort, undifferentiated rate of perceived exertion and subjective responses to written questions using either a 100 mm visual analogue scale or a free-format written response that inquired about practicality, comfort, adjustability, positive and negative feedbacks, and overall subjective preference in a simulated field trial (walking around a representative 'field-route' for 20 minutes at a self-selected pace). Perceived regional discomfort in the posterior neck area for the DSB (1.2, sd: 1.3) was significantly lower ($p<0.05$) than for the SSB (2.4, sd: 2.9). No significant differences were found in any of the other body regions. The DSB RPE (10.7, sd: 1.9) was significantly ($p<0.001$) lower during the last 30 seconds of walking than the SSB RPE (13.1, sd: 3.2). It was concluded that subjective perceptual methods are effective in distinguishing between single and double strap backpacks and that a DSB is superior to a SSB for student load carriage

Ling (2004) evaluated Modular Lightweight Load-Carrying Equipment (MOLLE) fitting on women while walking on level surfaces with different loads. Seven physically active women carried five levels of load (no load, 20, 30, 40 and 50 pounds) using MOLLE. With increased loads, women showed increased double-limb support time, decreased single-limb support time, increased trunk forward inclination excursion, decreased knee excursion, decreased medial-lateral excursion of center of gravity (COG), and increased vertical excursion of COG. Hip abductor strength was a strong predictor of COG vertical excursion. Some women required modification of the padded hip belt to ensure even weight distribution around the pelvis.

Perceptual responses did not distinguish between load carriage modes (Winsman and Goldman, 1976; Patton et al. 1990 and Kirk and Schneider, 1992). Winsman and Goldman (1976) used undifferentiated ratings of perceived exertion, in contrast, Legg and Mahanty (1985) reported significantly lower values for RPE when a load was carried using a combined front/backpack or a truncal jacket system than when the same load weight was carried in a backpack. However, they found no significant difference in RPE between a backpack with or without a frame. While Patton *et al.* (1990) used RPE's for the upper and lower body, Kirk and Schneider *(1992)* used RPE to distinguish between sensations in the chest, shoulders and legs.

7. Salient Findings from Review

The perusal of the literature in this chapter thus revealed the following:

1. The age and weight of the subject are the two most important factors which should be considered in ergonomic studies as physical activity and the muscle strength vary significantly with the age. The subjects aged between 20-35 years of age can be considered good for conducting the physiological studies.
2. The sample size taken for any ergonomic study should be such that the generalization about the population can be made in a more realistic way. If not taken into account, it will give faulty information about the population and also give misleading information about human capabilities and limitation.
3. Anthropometric data is most critical factor in designing the tools and equipment and significant differences were found with respect to race and other factors. If mismatch between human body dimensions and tool's dimensions occurs, it can cause various ailments and musculoskeletal disorders.
4. Heart rate and oxygen consumption rate are the two most important physiological parameters to study the human energy expenditures.
5. Modes of carrying load, surface grade, load (weight) and the environmental parameters (temperature and humidity)

are some of the factors which affect physiological cost of the worker.

6. Carrying load with a backpack affects human performance.

8. Conclusion

It can be concluded from the literature cited on various aspects of anthropometric data and carrying load, that there is scanty information available on the hill women. There is lack of anthropometric data on hill women in general and Uttarakhand in particular and also on carrying heavy loads uphill in different environmental conditions especially in India. Scarcity of information on above aspects provided a solid reason to undertake women as the subject to investigate effects of modes, grades, load and environmental conditions on physiological cost and perceived exertion as a research problem.

Chapter-3

Materials and Methods

In many ways the methodology of a study is more important than the analysis. A badly designed study can never be retrieved, whereas a poorly analyzed one can usually be reanalyzed. Consideration of design is also important because the design of a study will govern how the data are to be analyzed. The detailed sequential procedures for assessing the effect of load (weight), modes of carrying load, surface grade, and environmental parameters on physiological responses, perceived exertion and on body part discomfort are given below and consisted of eight sub-heads.

1. Research design
2. Theoretical framework for the study
3. Selection of variables
4. Operational definition
5. Selection of tool
6. Selection of sample
7. Method of data collection
8. Data analysis procedure

1. Research Design

"A research design is the arrangement of conditions for collection and analysis of the data in a manner that aims to combine relevance to research purpose with economy in procedure". A cross-sectional-cum-experimental design was used for present study.

2. Theoretical Framework for the Study

Manual material handling activities such as lifting, pulling, pushing and carrying of weights have been recognized as being associated with a high prevalence of musculoskeletal disorders in the working population. Most extensive studies and recommendation or legislation are on lifting while knowledge regarding carrying load by women is sparse. It is understood that the physiological responses, perceived exertion and body part discomfort of the subject (women) are affected by load (weight), modes of carrying load, surface grade, and environmental parameters and thus, one best method having least cost in terms of physiological responses, perceived exertion and body part discomfort will be selected and then recommendation on rest period will be given under different conditions and there after a back gadget will be developed for carrying load in hills. Theoretical framework for the study is given in the Fig. 3.1

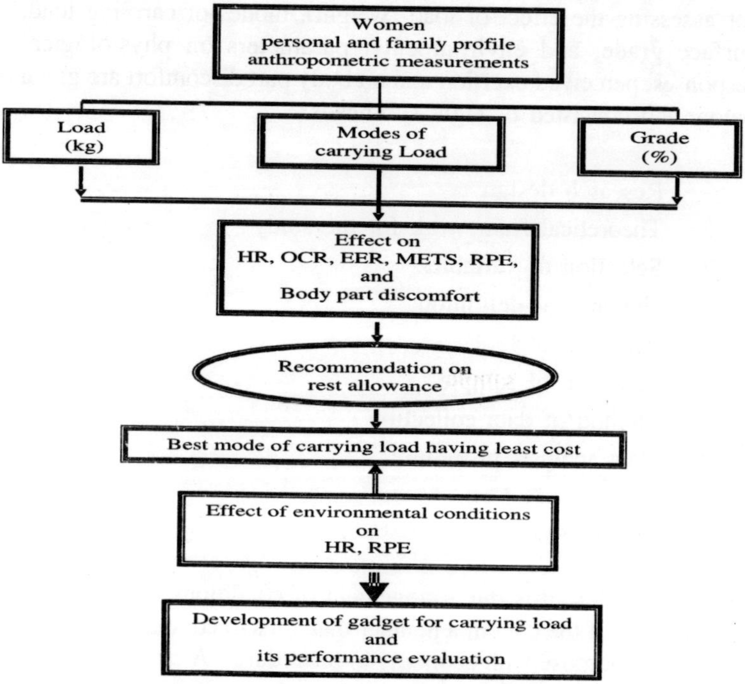

Fig. 3.1 : Theoretical frame work for the study

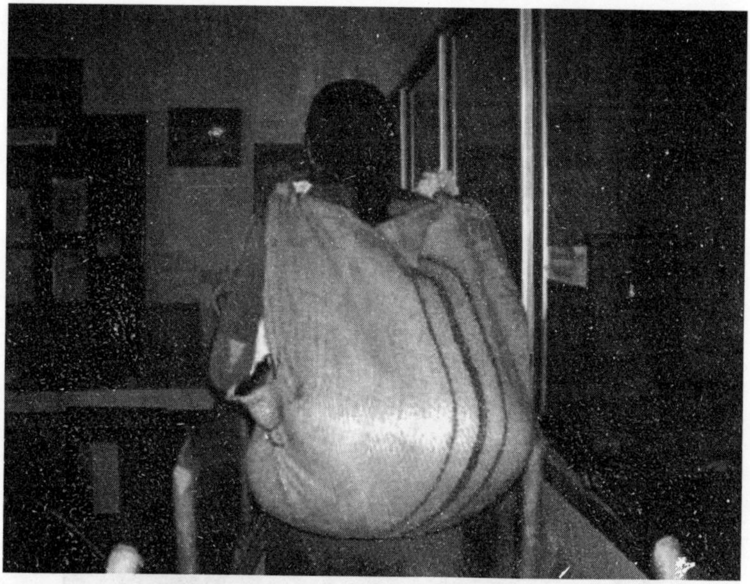

Plate 3.1 : *Carrying load on back (back view)*

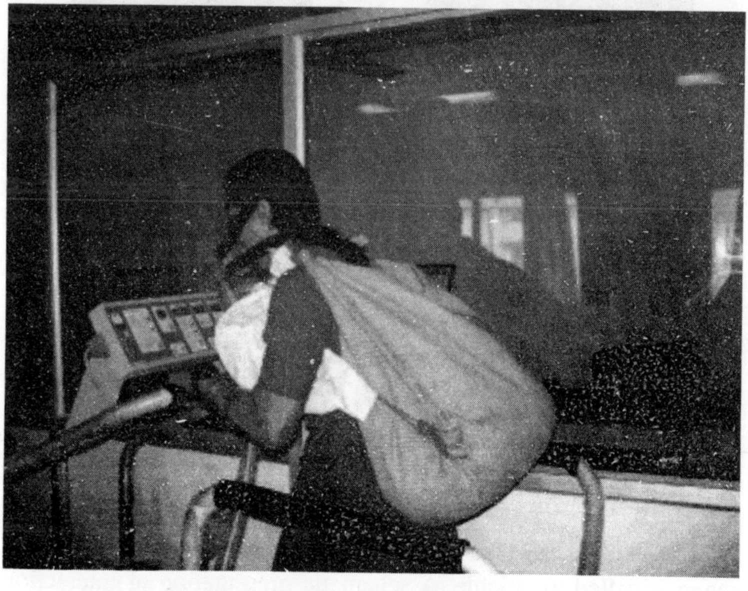

Plate 3.2 : *Carrying load on back (side view)*

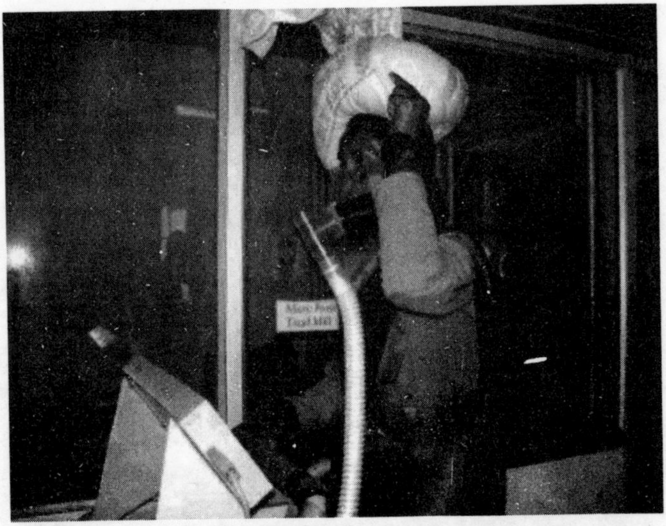

Plate 3.3 : *Carrying load on head*

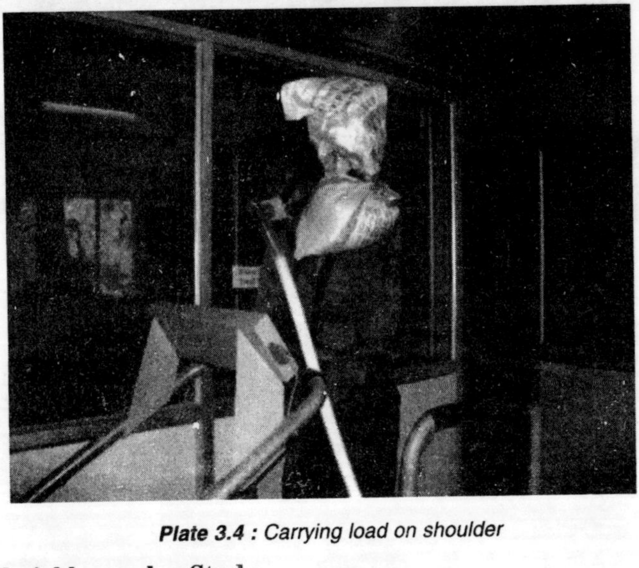

Plate 3.4 : *Carrying load on shoulder*

3. Variables under Study

A concept that can take on different quantitative value in a given range is called a variable. A schematic presentation of interaction of variable is given in Fig. 3.2.

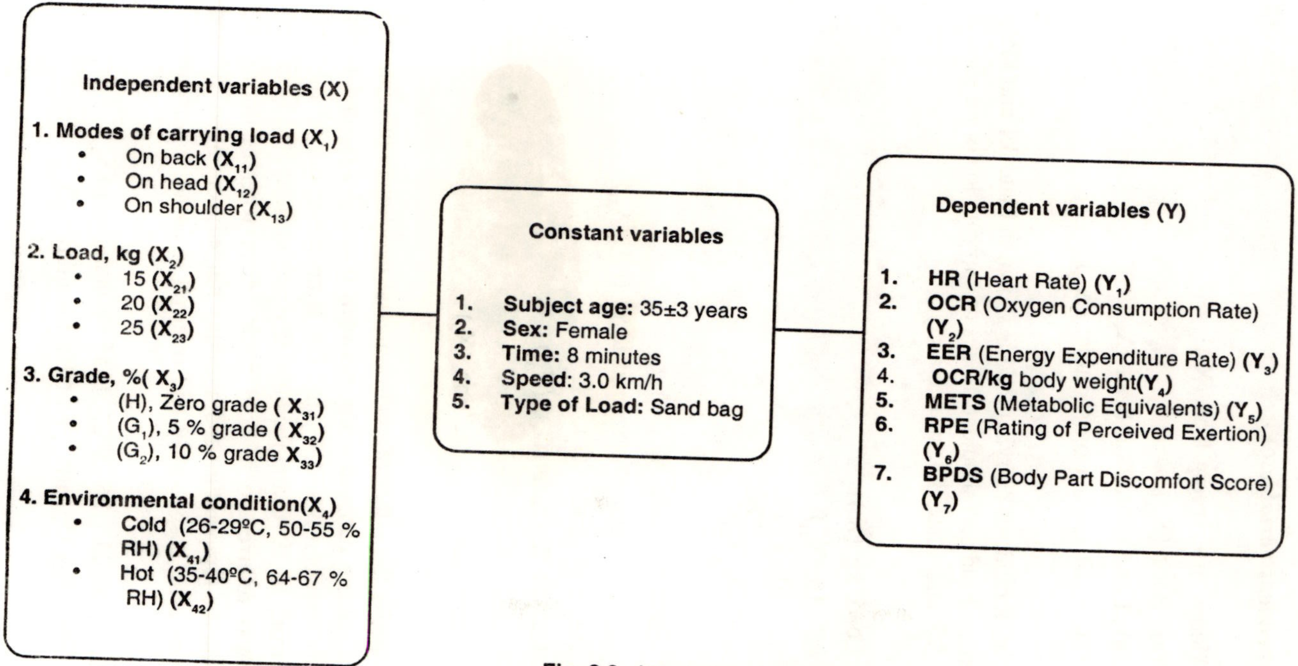

Fig. 3.2 : Interaction of variables

3.1 Independent variables(X)

(i) Modes of carrying load (X_1)

(a) Back ($\mathbf{X_{11}}$),

(b) Load carried on head ($\mathbf{X_{12}}$)

(c) Load carried on shoulder ($\mathbf{X_{13}}$)

The back mode of carrying load is shown in Plate 3.1 and Plate 3.2 where as head and shoulder modes of carrying loads are shown in Plate 3.3 and Plate 3.4, respectively.

(ii) Load, kg (X_2)

(a) 15 ($\mathbf{X_{21}}$)

(b) 20 ($\mathbf{X_{22}}$)

(c) 25 ($\mathbf{X_{23}}$)

(iii) Surface Grade, % (X_3)

(a) Zero grade/horizontal level (H), ($\mathbf{X_{31}}$)

(b) 5 % grade (G_1), ($\mathbf{X_{32}}$)

(c) 10 % grade (G_2), ($\mathbf{X_{33}}$)

(iv) Environmental condition (X_4)

To assess the effect of environmental parameters the data was recorded during early morning between 6 and 7 a.m. and in the afternoon between 3 and 4 p.m. The temperature during morning ranged from 26-29 °C and the humidity ranged from 50-55 per cent which was referred to as cold environment whereas during the afternoon the temperature and humidity ranged from 35-40 °C and 64-67 per cent, respectively and referred to as hot environment.

(a) Cold (26-29°C, 50-55 % humidity) ($\mathbf{X_{41}}$)

(b) Hot (35-40°C, 64-67% humidity). ($\mathbf{X_{42}}$)

3.2 Dependent variables(Y)

(i) Heart Rate (HR) beats/min (Y_1)
(ii) Oxygen Consumption Rate (VO_2) l/min (Y_2)
(iii) Energy Expenditure Rate (EER) kcal/min (Y_3)
(iv) Oxygen Consumption Rate/kg body weight (ml/min/kg) (Y_4)
(v) Metabolic Equivalents (METS) (Y_5)
(vi) Rating of Perceived Exertion (RPE) (Y_6)
(vii) Body Part Discomfort Score (BPDS) (Y_7)

3.3 Constant variables

(i) *Age and sex of the subject :*All subjects included in the experiments were women and from the same age group i.e. 35±3 years, without any history of chronic or acute illness, hypertension and also were not under any kind of medication and also were not pregnant.
(ii) *Time :* Time spent on each activity for each combination was same as 8 minutes.
(iii) *Speed of treadmill:* The speed was also kept constant i.e. 3.0 km/h for each combination.
(iv) *Type of load :* The load consisted of sand bag.

4. Operational Definitions

(i) *Anthropometric measurement:* It is the measurement of the human body, which includes body dimensions and the mechanical aspects of motion using anthropometric kit.
(ii) *Body Mass Index (BMI):* BMI (body weight in kg/height in m^2) is a simple and internationally used indicator in judging body weight and applies equally to men and women. A BMI of between 20 and 25 is considered to be the range with the lowest health risk. If the BMI is less than 18, it is underweight, a BMI of more than 25 is an indication that the subject is starting to be overweight and a BMI of over 30 is considered obese.

(iii) Body Surface Area (BSA): BSA is a measurement used in many medical tasks, such as for the determination of a proper dose of medication. Various formulas have been developed, all giving slightly different results. Turbofit software uses the Mosteller formula {BSA (m^2) = ([Ht (cm) x Wt (kg)]/3600)½}.

(iv) Carrying: It is the ability to hold materials in the hands or arms, on the shoulder or on the head while walking.

(v) Physiological responses: It includes the increased heart rate and energy expenditure of subjects while working.

(vi) Perceived exertion: It is the subjective evaluation of the exertion perceived by the subject on a scale of 0 (no exertion) to 10 (maximum exertion).

(vii) Physical Fitness Index (PFI): Physical fitness is the state of health of workers. To calculate PFI , the workers were asked to perform the exercise on step stool for 5 min then the heart rate, at the end of 1st, 2nd and 3rd minute of recovery was noted down and then PFI is calculated by the following formula:

$$PFI = \frac{\text{Duration of activity (seconds)}}{\text{Sum of 1st, 2nd and 3rd min.recovery of heart rate (beats/min)}} \times 100 \quad \text{..........3.1}$$

Subjects were ranked from poor to excellent on the basis of PFI as follows:

(a) Up to 80 – Poor physical fitness
(b) 81- 100 – Low average
(c) 101-115 – High average
(d) 116-135 – Good
(e) 136 -150 – Very good
(f) 151 and above – Excellent

(viii) Subject: The person on whom experiment is conducted.

(ix) VO_2 (Oxygen uptake): The energy yield from every liter of oxygen consumed by the body ranges from a low of 4.7

kcal/l with pure fat as the energy yielding substrate, to a high of 5.05 kcal/l, if the energy source is carbohydrate. The kcal yield per liter of oxygen uptake is taken as 4.85, if the person is utilizing the average mixed diet as the substrate pool. This condition would produce an R.Q. (respiratory quotient) value of 0.85. For a simple calculation of energy costs a value of 5.0 kcal/l of oxygen utilized is acceptable.

(x) *VO_2 max:* The predicted VO_2 max is based on a cross-section of normal, healthy individuals and is adjusted to age, weight, height, sex and form of exercise. The greatest amount of oxygen that can be taken up and used by the body for a given unit of time has been termed as the maximal oxygen uptake (VO_2 max) or aerobic capacity.

(xi) *METS:* A MET equals 3.5 milliliter of VO_2 per kilogram of body weight.

5. Selection of Tool

- The interview schedule was developed for collecting the data on personal, family, musculoskeletal disorders, injuries, slips, trips, fall and medical history of the subjects.
- For anthropometric data base, data table was developed.
- For physiological responses, data tables were developed
- *Borg General Scale (1982)* was used for assessing the perceived exertion of subject while performing various load carrying activities (Fig. 3.3).

Instructions to use Borg's scale

1. I will ask you to specify a feeling but do select a number which most accurately suits your perception of fatigue/ exertion.
2. If you do not feel anything or if there is no symptom, answer is zero.

0.0	Nothing at all
0.5	Extremely weak
1.0	Very weak
2.0	Weak
3.0	Moderate
4.0	Some what strong
5.0	Heavy
6.0	Strong
7.0	Very strong
8.0	Extremely strong
9.0	Almost maximum
10.0	Maximum

Fig. 3.3 : Borg general 10 point scale (1982)

3. If you start feeling noticeable, your answer is 0.5 i.e. extremely weak, likewise if you have a strong feeling answer will be 6 and so on.
4. More you feel the stronger the feeling, higher the number you choose.
5. Keep in mind, there is no wrong numbers, be honest and do not over react and underestimate your rating.
6. Do you have any questions?

- For body part discomfort, Corlett and Bishop (1976) scale was used to measure local muscle pain and discomfort. Body part discomfort score (BPDS) is the measure of localized discomfort which may restrict the subject to continue the work.

Various equipment were also used for collecting the data which included the following:

(*a*) *Anthropometric kit:* Anthropometric kit was used to measure the dimensions of various body parts of women.

(*b*) *Measuring tape:* Measuring tape was used to measure waist and chest circumferences.

(*c*) *Weighing Balance:* Weighing balance was used to measure the weight of the women and each time it was adjusted to its zero point.

(*d*) *Spreading caliper:* Spreading caliper was used to measure the head breadth and head length.

(*e*) *Polar heart rate monitor:* Heart rate was measured by polar heart rate monitor. The machine consisted of chest belt transmitter, elastic strap, wrist receiver unit and interface. The electrode in the transmitter senses the heart beat and transmits these signals in digital form or beats per minute (beats/min). The receiver displays these signals in digital form. For taking data, the backside of the transmitter was made wet with water and attached to the chest belt and worn around the chest of the subject just at the centre of lungs.

(*f*) *Vista VO_2 lab:* The vista VO_2 lab is a high precision instrument comprising of O_2 analyzer, turbine flow sensor, barometric pressure and temperature sensor. The analyzer uses a NAFION gas dryer/humidifier. This consisted of a water vapour permeable tube that absorbs moisture and releases this excess moisture to atmosphere. Therefore, moisture is reduced to a level of ambient humidity. Conversely; a dry calibration gas is humidified to the level of ambient humidity. Holding treadmill handrails will cause an underestimation of VO_2 of up to 15 per cent. The complete set of equipment is shown in Plate 3.5.

(*g*) *Treadmill:* It is model number S787 with 887S programmable treadmill controller, a motorized jogging machine which goes through various speeds and elevations. The S- series model has wider belt for safety. It is 1955 (77") mm long and 711(28") mm wide.

(*h*) *Thermo hygrometer:* It was used to measure temperature and humidity. The user can view maximum and minimum temperature/humidity by pressing maximum and minimum key.

(i) *Lux meter:* It was used to measure the intensity of light.

(j) *Sound level meter:* It was used to measure the sound level.

6. Selection of Sample

6.1 Selection of locale

Uttarakhand state {Tehri (Garhwal) and Nainital (Kumaon)} was purposively selected for the study.

6.2 Sampling design

- For anthropometric data simple random sampling without replacement was used to select the sample (Fig. 3.4)

Plate 3.5 : Complete set of Vista VO_2 lab

- For laboratory experiment the subjects of same age group were selected by non-probability sampling from U.S. Nagar and Nainital districts.

6.3 Sample size

- ū For anthropometric data, 100 women were selected of 20-45 years of age group.
- ū For experimental data, 7 women subjects of age group 35±3 years were selected on the basis that they are not having any history of chronic illness and are not currently under any medication that would alter cardiovascular or metabolic function. Prior consent of each subject was taken. The information regarding the kind of experiment was given to each subject.

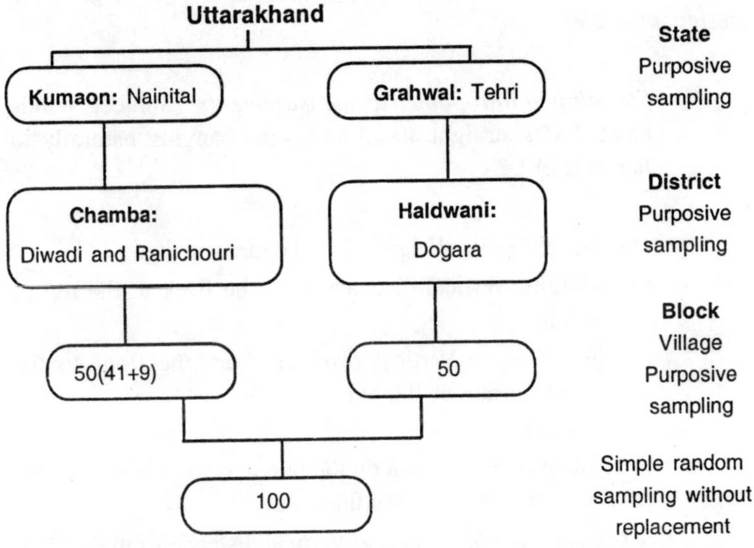

Fig. 3.4 : Sampling design for anthropometric data

7. Data Collection Procedure

Phase 1

Anthropometric data of Uttarakhand women was collected. A total of 42 body measurements were taken which were categorized under five categories i.e. 7 standing anthropometric measurements, 10 sitting anthropometric measurements, 7 anthropometric measurements of hands, 10 body breadths and 8 body circumferences and depths.

Subjects were measured in both standing and sitting postures. All subjects were clothed as lightly as possible and were barefoot. Standing measurements were taken in an erect, but not rigid posture. The sitting posture was arranged so that the feet rested flat on the floor, and an angle of 90° was formed at the knee and the hip joint. Definitions of the all anthropometric measurements were adopted from the book *'Human Engineering Guide to Equipment Design'* besides the chest and waist circumferences which were taken according to *'Indian Anthropometric Dimensions for ergonomic design practice'*.

(i) *Standing anthropometric measurements* (Subject stands erect, looks straight ahead and arms hanging naturally to her sides.)

- *(a)* *Weight (kg):* Weight of body mass in kg.
- *(b)* *Height:* Vertical distance from the floor to the top of head.
- *(c)* *Eye height:* Vertical distance from the floor to the inner corner of the eye.
- *(d)* *Shoulder height:* Vertical distance from the floor to the uppermost point on the lateral edge of the shoulder with the subject standing straight.
- *(e)* *Elbow height standing:* Vertical distance from the floor to the depression at the elbow between the bones of the upper arm and forearm.
- *(f)* *Knuckle height standing:* Vertical distance from the floor to the point where the middle finger of the right hand meets palm.
- *(g)* *Overhead reach:* Measurement is made when subject stands erect, grasp a horizontal bar above her right

side with her right hand (at the bend formed by palms and fingers), and raises bar to the highest position attainable without strain.

(ii) *Sitting anthropometric measurements* (Subject sits erect on chair and looks straight ahead)

(a) *Sitting height:* Vertical distance from sitting surface to the top of head.

(b) *Eye height sitting:* Vertical distance from seat surface to the inner corner of eye.

(c) *Shoulder height sitting:* Vertical distance from seat surface to the uppermost point on the lateral edge of the shoulder

(d) *Elbow height sitting:* Vertical distance from seat surface to the bottom of the right elbow. The subject sits erect with her upper right arm vertical at her side and her forearm at a right angle to the upper arm.

(e) *Thigh height sitting:* Vertical distance from seat surface to the top of the thigh at its intersection with the abdomen. The subject sits erect with her knees and ankles at right angles.

(f) *Knee height sitting:* Vertical distance from the floor to the uppermost point on the knees. The subject sits erect with her knees and ankles at right angles.

(g) *Poplietal height sitting:* Vertical distance from the floor to the underside of the thigh immediately behind the knee. The subject sits erect with her knees and ankles at the right angles and the bottom of her thighs and the back of her knees barely touching the sitting surface.

(h) *Buttock-leg length:* Horizontal distance from the most posterior point on the buttocks to the base of the heel. The subject sits erect with her legs as far as possible on a horizontal surface.

i. *Buttock-knee length:* Horizontal distance from the most posterior point on the buttocks to the most anterior point on the knee. The subject sits erect with her knees and ankles at right angles.

j. *Buttock-poplietal length:* Horizontal distance from the most posterior point on the buttocks to the back of the lower leg at the knee. The subject sits erect with her knees and ankles at the right angles.

(iii) Anthropometric measurements of hands

(a) *Hand breadth at metacarpal:* Maximum breadth across the end of the metacarpal bone (where the fingers join the palm) of the index and little finger of the right hand held straight and stiff with the fingers together.

(b) *Hand breadth at thumb:* Maximum breadth across the palm (at right angles to the long axis of the hand) at the knuckle of the thumb of the right hand with the fingers extended and the thumb lying alongside and in the plane of the hand.

(c) *Hand length:* Distance from the base of the thumb to the middle fingertip of the right hand held straight and stiff.

(d) *Hand thickness:* Maximum distance between the dorsal and palmar surface of the knuckle of the middle finger where it joins the palm of the right hand when the fingers are extended.

(e) *Shoulder-elbow length:* Vertical distance from the uppermost point on the lateral edge of the shoulder to the bottom of the elbow.

(f) *Forearm-hand length:* Horizontal distance from the tip of the right elbow to the tip of the longest finger. The subject sits erect with her upper right arm vertical at her side and her forearm, hands and fingers extended horizontally.

(g) *Arm length:* Horizontal distance from the uppermost

point on the lateral edge of the shoulder to the tip of the extended middle finger.

(iv) Body breadths

(a) Interpupillary distance: Horizontal distance between the centres of the pupils with the subject looking straight ahead.

(b) Head length: Distance between the most anterior point on the head (between the brow ridges) and the most posterior point on the head in the midplane of the body.

(c) Head breadth: Maximum head breadth above the ears and at the right angles to the midplane of the body.

(d) Shoulder breadth: Maximum horizontal distance across the deltoid muscles.

(e) Chest breadth: Horizontal distance across the chest at nipple level.

(f) Maximum body breadth: Maximum breadth of the body, including arms, as the subject stands erect with her arms hanging relaxed at her side.

(g) Elbow-elbow breadth: Maximum horizontal distance across the lateral surface of the elbows. The subject sits erect with her upper arms vertical and touching her sides and her forearms extended horizontally.

(h) Hip breadth standing: Maximum horizontal distance across the hips. The subject stands erect with her heels together.

(i) Hip breadth sitting: Maximum horizontal distance across the hips. The subject sits erect with her knees and ankles at the right angles and her knees and heels together.

(j) Knee-knee breadth: Maximum horizontal distance across the lateral surfaces of the knees. The subject sits erect with her knees at right angles and pressed together.

(v) *Body circumferences and depths*

(a) *Chest circumference:* It is measured on the chest at nipple level.

(b) *Waist circumference:* It is measured on the waist, at the level of the upper margin of the lateral iliac crests, (where the belt is worn).

(c) *Chest depth:* Horizontal distance from the front to the back of the chest at nipple level (on women, at the level where 4th rib meets the breastbone).

(d) *Waist depth (Standing):* Horizontal distance between the back and the abdomen at the level of the greatest lateral indentation of the waist (if this is not apparent, at the level at which the belt is worn).

(e) *Maximum body depth:* Maximum horizontal distance between the vertical planes passing through the most anterior points on the trunk.

(f) *Buttock depth:* Horizontal distance between the buttocks and the abdomen at the level of the maximum protrusion of the buttocks.

(g) *Foot breadth:* Maximum horizontal distance across the foot at right angles to the long axis. The subject stands with her weight equally distributed on both feet.

(h) *Foot length:* Horizontal distance from the heel to the tip of the longest toe.

Phase 2

In this phase laboratory experiment was conducted and data on physiological responses, perceived exertion and body part discomfort were taken.

The following *test protocol* was followed for conducting laboratory experiment:

(i) Orientation of the subjects with treadmill walking and other equipment was done before conducting the experiment so that they do not feel any problem during experiment. Also, they were told not to talk while conducting the experiment and only to use hand signals.

(ii) Before start of the experiment, subject was allowed to rest for a minimum period of 30 minutes.

(iii) Polar heart rate transmitter belt was affixed on the chest of the subject and wrist receiver was fixed on the hand of the subject.

(iv) The disposable air cushion face mask (Q-valve) was fitted to the subject such that the head gear straps were fairly tight so that no amount of air should leak. A separate facemask was used for each individual.

(v) The subject was allowed to take five minutes of rest; meanwhile the subject's personal data was entered in to the software after selecting the stress test.

(vi) The speed of treadmill and time for each treatment combination was kept constant i.e. 3.0 km/h and 8 minutes, respectively.

(vii) The subject was asked to walk on the treadmill on the endless belt of treadmill. The subjects were allowed to hold the handrails. The heart rate (from 5th minute of walking till 8th minute for each minute) and oxygen consumption (data reporting interval every 5 breaths with a 1 breath average till 8th minute) were recorded.

(viii) Then subject was allowed to take rest till their physiological parameters came to normal or at resting level after each combination trial and this time was recorded as to be recommended for rest allowance in load carrying activities by hill women and then the subject was given water.

(ix) The step *(viii)* was followed for each combination for recommendation on rest allowances.

(x) For each combination seven replications were used.

Phase 3

To study the effect of environmental parameters (temperature and humidity) the heart rate (from 5th minute of walking till 8th minute for each minute) and perceived exertion were recorded for carrying load on back at horizontal level.

Phase 4

In this phase the data was collected on the performance evaluation of the especially designed external frame backpack for hill women in laboratory and field conditions with three replications.

8. Data Analysis Procedure

Basis of data categorization

- The categorization of anthropometric data was done on the basis of Garhwal and Kumaon region of Uttarakhand state.
- The physiological responses, perceived exertion and body part discomfort were categorized on the basis of the independent variables.

Statistical analysis

(i) In the present study frequency, percentage, mean, standard deviation and percentile (5th, 50th, and 95th) were used to analyze the anthropometric data.

(ii) Hypothesis testing:

(a) The mean anthropometric measurements of Garhwal and Kumaon regions are equal.

(b) The effect of various modes of carrying load, load and grade on HR, increase in HR, OCR, EER, OCR/kg body weight, METS and RPE is same.

(c) There exists a relationship between (HR and OCR), (HR and RPE) and (OCR and RPE).

(d) The HR and OCR are a function of mode, load and grade.

- For testing hypothesis (a) analysis of variance was used.
- Three-factorial CRD was used for testing (b) hypothesis.
- Correlation coefficient and regression analysis were used for testing (c) hypothesis.
- Multiple regression analysis was used for testing (d) hypothesis.

Chapter-4

Results and Discussion

The results of the present study are described and discussed in this chapter. The entire descriptive part has been described on the basis of Garhwal and Kumaon regions of Uttarakhand whereas the experimental part was described on the basis of modes of carrying load, grade, amount of load and environmental parameters. The findings of the study were produced through composite summary tables followed by statistical application to find out the significance level and relational statistics among variables. The results are presented under the following sections:

1. Demographic profile of hill women of Uttarakhand
2. Medical history of hill women of Uttarakhand
3. Pain in various body parts of hill women of Uttarakhand during carrying loads in last six months
4. 'Physical Fitness Index (PFI)' of hill women of Uttarakhand
5. Anthropometric measurements of hill women of Uttarakhand
6. Comparison of anthropometric measurements of hill women of Uttarakhand with women of other states
7. Demographic profile of women subjects employed for experiments
8. Anthropometric measurements and Physical Fitness Index (PFI) of women subjects employed for laboratory experiments

9. Environmental conditions during laboratory experiments
10. Effect of interaction of load, mode and grade on physiological responses and perceived exertion
11. Effect of interaction of main factors i.e. (load and mode), (load and grade) and (mode and grade) on physiological responses and perceived exertion
12. Effect of main factors i.e. loads, modes and grades on physiological responses and perceived exertion
13. Correlation and regression analysis between various parameters
14. Regression analysis between heart rate and independent variables (grade, mode and load)
15. Regression analysis between oxygen consumption rate and independent variables (grade, mode and load)
16. Recommendation on rest periods for women subjects during various modes of carrying loads on different surface grades
17. BPDS (Body Part Discomfort Score) for different modes of carrying loads on various grades
18. Effect of different environmental conditions (temperature and humidity) on HR and RPE for back mode of carrying load on horizontal plane
19. Development of the backpack for carrying load in hills and its performance evaluation
20. Performance evaluation of developed external frame backpack

1. Demographic Profile of Hill Women of Uttarakhand

Table 4.1 shows the demographic profile of hill women of Uttarakhand which has been described as follows:

Mean age

The mean age for Garhwal and Kumaon regions varied as 34.9 and 36.82 years, respectively in the age group of 20-45 years. The

Table 4.1: Demographic profile of hill women of Uttarakhand

S. No.	Demographic characteristics	Garhwal N=50	Kumaon N=50	Uttarakhand N=100
1.	**Mean age (Years)**	34.9	36.82	35.86
2.	**Education**		Frequency	
	(i) Illiterate	6(12)	4(8)	10(10)
	(ii) Primary	15(30)	11(22)	26(26)
	(iii) Junior high school	11(22)	8(16)	19(29)
	(iv) High school	8(16)	13(26)	21(21)
	(v) Intermediate	9(18)	10(20)	19(9)
	(vi) Graduate	2(4)	3(6)	5(5)
	(vii) Post-graduate	–	–	–
3.	**Family type**			
	(i) Nuclear	28(56)	24(48)	52(52)
	(ii) Joint	15(30)	20(40)	35(35)
	(iii) Extended	7(14)	6(12)	13(13)
4.	**Family-size**			
	(i) 1-3 (Small)	4(8)	3(6)	7(7)
	(ii) 4-7 (Medium)	32(64)	36(72)	68(68)
	(iii) > 8 (Large)	14(28)	11(22)	25(25)
5.	**Source of income**			
	(i) Agriculture	50(100)	50(100)	100(100)
	(ii) Business	6(12)	5(10)	11(11)
	(iii) Service	15(30)	8(16)	23(23)
	(iv) Livestock (Dairy)	50(100)	47(94)	97(97)
6.	**Monthly income (Rs)**			
	(i) 500-1000	11(22)	12(24)	23(23)
	(ii) 1001-2000	14(28)	16(32)	30(30)
	(iii) 2001-3000	7(14)	8(16)	15(15)
	(iv) 3001-4000	4(8)	4(8)	8(8)
	(v) 4001-5000	9(18)	7(14)	16(16)
	(vi) >5000	5(10)	3(6)	8(8)
7.	**Average time spent (h/day)**			
	(i) Collecting fuel, fodder etc	3-5	2-3	3.5
	(ii) Agricultural activity	5-6	6-8	6.8

Note: Figures in parentheses indicate percentage of entire sample.
N = Sample size

over- all mean age of the selected hill women of Uttarakhand was found as 35.86 years.

Educational qualification

Out of total sample 10 per cent of women in the age group of 20-45 years were illiterate whereas 26 per cent were having primary education and 19 per cent were in 5th - 8th class pass category. About 21 and 19 per cent of women were high school and intermediate, respectively and only 5 per cent of them were graduates and were either newly married women or girls in the age group of 20-25 years. Out of the total, about 4 per cent graduates were from Garhwal and 6 per cent were from Kumaon region. One of the main reasons for higher percentage of high school and intermediate women from Kumaon region might be the proximity of their habitate to a town named as Haldwani where adequate educational facilities are available.

Family type

About 56 per cent from Garhwal and 48 per cent from Kumaon belonged to nuclear family type, thus nuclear type of family was predominant in which 52 per cent belonged to this category; whereas 35 and 13 per cent were living in joint and extended types of families, respectively.

Family size

The large, medium and small size families were noted as 25, 68 and 7 per cent, respectively. In Kumaon region, 72 per cent of the families were of medium size whereas in Garhwal region it was about 64 per cent. The small size families were only 6 and 8 per cent in Kumaon and Garhwal region, respectively.

Source of income

Agriculture and livestock are the two main sources of income and almost 100 per cent of them are earning their livelihood from these two sources whereas some of the families i.e. 11 per cent were having business in the form of shops, flour mill and PCOs, and about 23 per cent were in service. About 30 per cent women's family member from Garhwal region and 16 per cent from Kumaon region were in service outside their villages. The results show that still in hill regions of Uttarakhand, agriculture and livestock are the major sources of income.

Monthly income

The monthly income of the families was found very low. About 23 per cent families were having monthly income in the range of 500-1000 Rs/month, about 16 per cent of families were having income in range of 4000-5000 Rs/month and only about 8 per cent of families were having income above Rs 5000 Rs/month. Families with higher income were those families where family members were earning through service outside village or having shops etc. About 30 per cent and 15 per cent of families were having income in between 1001-2000 Rs/month and 2001-3000 Rs/month, respectively. This reveals that still in the hill regions of Uttarakhand monthly income of the families was very low due to lower productivity from agriculture and livestock. The returns from these sources were not in tune with the amount of effort and time put up by hill women of Uttarakhand.

Average time spent in collecting fuel, fodder etc. and agricultural activities

It was found that in Garhwal region the women on an average spent 3-5 h/day in collecting fuel/fodder whereas 5-6 h/day in agricultural activities. Similarly, in Kumaon region average time spent was about 2-3 h/day in collecting fuel/fodder and about 6-8 h/day in agricultural activities. It could be therefore, inferred that in general women spent 3.5 and 6.8 h/day in collecting fuel/fodder etc. and in agricultural activities, respectively. Moreover, women of Garhwal region spent more time in collecting fuel/fodder and less time in agricultural activities than women of Kumaon region.

2. Medical History of Hill Women of Uttarakhand

Table 4.2 shows the medical history of hill women of Uttarakhand during the last six months and their various illnesses are mentioned below:

Table 4.2: Medical history of hill women of Uttarakhand

S. No.	*Illness*	*Garhwal N=50*	*Kumaon N=50*	*Uttarakhand N=100*
1.	**Common Illness**			
	(i) Cold, cough	15(30)	18(36)	33(33)
	(ii) Headache,fever	25(50)	28(56)	43(43)
	(iii) Digestive problems	14(28)	12(24)	26(26)
	(iv) Hearing and sight problems	8(16)	6(12)	14(14)
2.	**Chronic Illness**			
	(i) Diabetes	1(2)	–	1(1)
	(ii) Blood pressure	3(6)	4(8)	7(7)
	(iii) Respiratory	4(8)	2(4)	6(6)
3.	**Deficiency diseases**			
	(i) Fe-deficiency	–	–	–
	(ii) Ca-deficiency	–	–	–
	(iii) Any other	–	–	–

Note: *Figures in parentheses indicate percentage of entire sample.*
N= Sample size

Common illnesses

It was found that about 33 per cent suffered from common illnesses such as cold and coughs whereas about 43 per cent suffered from headaches and fevers and about 26 and 14 per cent suffered from digestive problems, and hearing and sight problems, respectively.

Chronic illnesses

It was found that only 1 per cent from the total sample was suffering from diabetic who belonged to Garhwal whereas 7 and 6 per cent of total sample suffered from blood pressure and respiratory diseases, respectively.

Deficiency diseases

None of the women from hill region of Uttarakhand were found to have any kind of the deficiency diseases as per their statements.

3. Pain in Body Parts while Carrying Loads

Table 4.3 showed the pain in various body parts of the hill women of Uttarakhand while carrying loads in last six months whereas Figs. 4.1 and 4.2 depicts the pain felt by Uttarakhand women, and Garhwal as well as Kumaon women, respectively. It was found that the back was the most badly affected part of the body, and about 79 per cent women felt pain in the back while carrying loads followed by the body pain (70 per cent), and shoulder pain (46 per cent). The neck pain and forearm pain were perceived by 28 and 26 per cent of women, respectively. About 31, 20, 18 and 15 per cent women felt pain in knee, foot, palms and biceps, respectively.

About 82 per cent of women slipped while carrying heavy load on the graded hill surface whereas about 67 per cent fell while carrying loads. About 46 per cent women had cuts and wounds

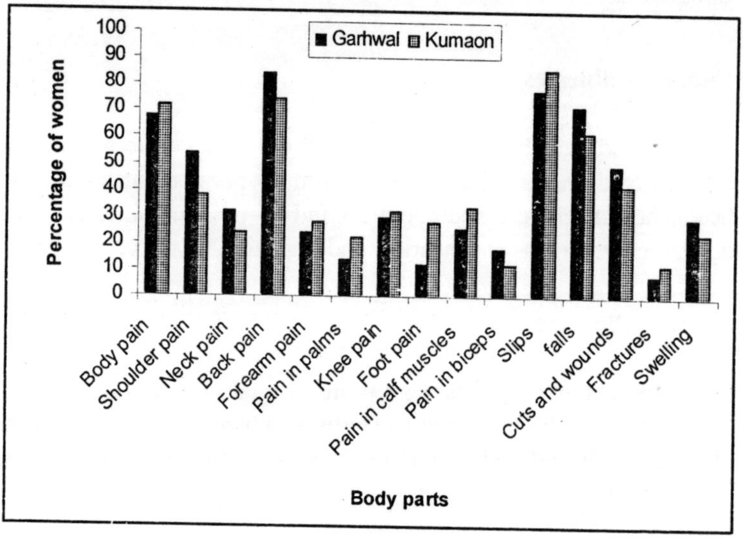

Fig. 4.1 : Pain in body parts of Garhwal and Kumaon women while carrying loads

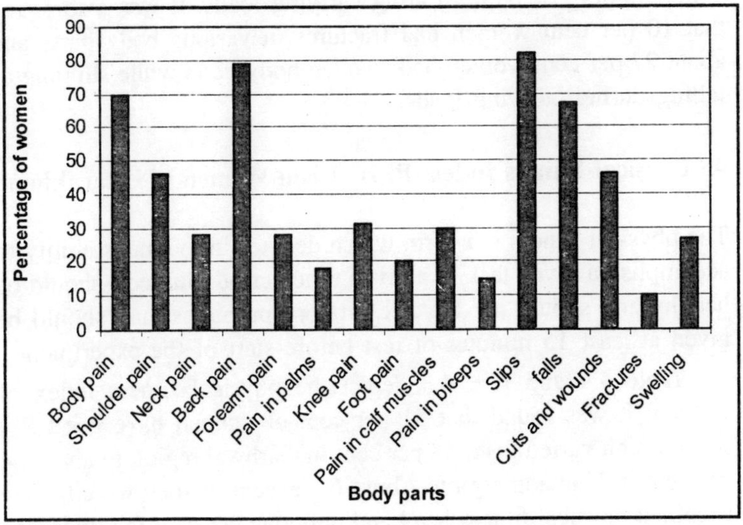

Fig. 4.2 : Pain in body parts of Uttarakhand women while carrying loads

Table 4.3 : Pain in various body parts of hill women of Uttarakhand during carrying loads

S. No.	Body parts	Frequency		
		Garhwal N=50	Kumaon N=50	Uttarakhand N=100
(i)	Body pain	34(68)	36(72)	70(70)
(ii)	Shoulder pain	27 (54)	19(38)	46(46)
(iii)	Neck pain	16(32)	12(24)	28(28)
(iv)	Back pain	42(84)	37(74)	79(79)
(v)	Forearm pain	12(24)	14(28)	26(26)
(vi)	Pain in palms	7(14)	11(22)	18(18)
(vii)	Knee pain	15(30)	16(32)	31(31)
(viii)	Foot pain	6(12)	14(28)	20(20)
(ix)	Pain in calf muscles	13(26)	17(34)	30(30)
(x)	Pain in biceps	9(18)	6(12)	15(15)
(xi)	Slips	39(78)	43(86)	82(82)
(xii)	Falls	36(72)	31(62)	67(67)
(xiii)	Cuts, wounds and injuries	25(50)	21(42)	46(46)
(xiv)	Fractures	4(8)	6(12)	10(10)
(xv)	Swelling	15(30)	12(24)	27(27)

Note: Figures in parentheses indicate percentage of entire sample.
N= Sample size

while slipping or falling during carrying loads. It was also found that 10 per cent women had fractures in various body parts and about 27 per cent women had swollen body parts while slipping or falling during carrying loads.

4. Physical Fitness Index (PFI) of Hill Women of Uttarakhand

The physical fitness is a term which denotes individual's ability to accomplish a given task in a given time period. Subjects should be healthy and should not have respiratory problems, and should be given at least 15 minutes of rest before start of the experiment.

Table 4.4 and Fig. 4.3 depict the physical fitness index of women. It was found that 51 per cent of women have good PFI score which varied from 48 per cent in Garhwal region to about 54 per cent in Kumaon region. About 6 per cent women were having excellent physical fitness level and only one per cent fell under the category of low average physical fitness. About 29 and 13 per cent of women were having high-average and very good physical fitness level, respectively and none of the women from selected sample were having poor physical fitness level. It could, therefore, be concluded that most of them were having high average to very good level of physical fitness.

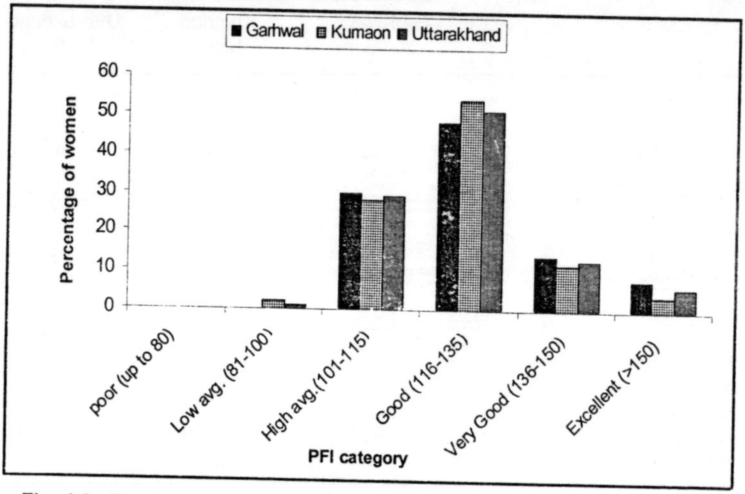

Fig. 4.3 : Distribution of Uttarakhand women on the basis of Physical Fitness Index (PFI)

Table 4.4: Physical Fitness Index (PFI) of hill women of Uttarakhand

S. No.	*Physical Fitness Index (PFI)*	*Frequency*		
		Garhwal N= 50	*Kumaon* N=50	*Uttarakhand* N=100
1.	Poor (Up to 80)	-	-	-
2.	Low average (81-100)	-	1(2)	1(1)
3.	High average (101-115)	15(30)	14(28)	29(29)
4.	Good (116-135)	24(48)	27(54)	51(51)
5.	Very good (136-150)	7(14)	6(12)	13(13)
6.	Excellent (Beyond 150)	4(8)	2(4)	6(6)

Note: Figures in parentheses indicate percentage of entire sample.
N= Sample size

$$PFI = \frac{\text{Duration of activity (minute)}}{\text{Sum of 1}^{st}\text{, 2}^{nd}\text{ and 3}^{rd}\text{ minute recovery heart rate (beats/min)}} \times 100$$

5. Anthropometric Measurements of Hill Women of Uttarakhand

A total of 42 body measurements were taken, which were grouped under five categories, viz. 7 standing anthropometric measurements, 10 sitting anthropometric measurements,7 anthropometric measurements of hands, 10 body breadths and 8 body circumferences and depths. Figs. 4.4 and 4.5 depict the line diagrams illustration of Uttarakhand women (20-45 years) in standing and sitting postures based on their mean anthropometric measurements. Analysis of variance is given in Appendix A, Table A-1.

Standing anthropometric measurements

Table 4.5 shows the standing anthropometric measurements (F-values, mean and standard deviation) and Table 4.5a shows various percentiles values (5th, 50th and 95th).

It was found that the mean weight of Uttarakhand women was 40.27 (±5.08) kg which varied as 40.8 (±4.58) kg and 39.74 (±5.48) kg for Garhwal and Kumaon regions, respectively. No significant difference was found in weight between women of Garhwal and Kumaon regions.

It was found that mean height, eye height, shoulder height,

elbow height standing, knuckle height standing and overhead reach of Uttarakhand women were 151.49 (±5.93), 139.79(±7.09), 126 (±5.12), 97.16 (±5.23), 67.41 (±7.01) and 167.5 (±6.33) cm, respectively. Significant difference was found only in case of eye height ($p<0.05$) between women of Garhwal and Kumaon regions.

Table 4.5 : Standing anthropometric measurements of Uttarakhand women

S. No.	*Standing anthropometric measurements (cm)*	*F-values*	*Mean*		
			Garwhal	*Kumaon*	*Uttarakhand*
1.	Weight (kg)	1.086 ns	40.80 (4.58)	39.74 (5.48)	40.27 (5.08)
2.	Height	0.007 ns	151.48 (4.43)	151.6 (6.98)	151.49 (5.93)
3.	Eye height*	5.101**	138.22 (6.25)	141.36 (7.51)	139.79 (7.09)
4.	Shoulder height	0.608 ns	126.4 (4.89)	125.6 (5.30)	126.00 (5.12)
5.	Elbow height	2.0618 ns	96.32 (5.60)	98.00 (4.69)	97.16 (5.23)
6.	Knuckle height	0.034 ns	67.54 (8.40)	67.28 (5.31)	67.41 (7.01)
7.	Overhead reach	1.0521 ns	166.72 (4.40)	168.28 (7.69)	167.50 (6.33)

Note: ** = Significant at 1 per cent level of significance,
* = Significant at 5 per cent level of significance
Values in parentheses indicate the standard deviation of Mean

Table 4.5a: Percentile values of standing anthropometric measurements of Uttarakhand women

S. No.	Standing anthropometric measurements (cm)	Percentile								
		5th	50th	95th	5th	50th	95th	5th	50th	95th
1.	Weight (kg)	33.4	40.0	46.55	32.0	38.5	48.0	32.95	40.0	48.0
2.	Height	145.0	151.5	159.1	139.4	152.0	164.2	141.9	152.0	161.0
3.	Eye height*	127.4	138.0	148.5	127.9	142.0	155.7	127.0	140.0	149.1
4.	Shoulder height	119.0	126.0	135.0	117.9	125.0	133.5	119.0	125.0	135.0
5.	Elbow height	90.0	97.0	104.0	89.4	98.0	104.0	89.9	98.0	104.0
6.	Knuckle height	58.9	66.0	83.2	58.0	67.5	75.0	58.0	67.0	76.0
7.	Overhead reach	159.4	168.0	172.0	157.4	169.0	179.5	158.0	168.0	179.0

Note: ** = Significant at 1 per cent level of significance,
* = Significant at 5 per cent level of significance
Values in parentheses indicate the standard deviation of Mean

Sitting anthropometric measurements

Table 4.6 shows the sitting anthropometric measurements (F-values, mean and standard deviation) and Table 4.6 shows various percentiles values (5th, 50th and 95th).

It was found that the mean sitting height, eye height, shoulder height, elbow height, thigh height, knee height, poplietal height of Uttarakhand women were 74.96 (±3.68), 64.40 (±3.89), 50.89 (±3.70), 18.96 (±2.75), 11.80 (±1.71), 46.98 (±4.64), 41.28 (±3.87) cm, respectively. The buttock – leg length, buttock – knee length and buttock – poplietal length were 95.57 (±5.88), 49.55 (±4.11) and 43.28 (±3.95) cm, respectively. Significant differences were found between Garhwal and Kumaon regions in sitting height ($p<0.05$) and sitting elbow height ($p<0.01$).

Table 4.6 : Sitting anthropometric measurements of Uttarakhand women

S. No.	*Sitting anthropometric measurements (cm)*	*F-value*	*Mean*		
			Garhwal	*Kumaon*	*Uttarakhand*
1.	Sitting height*	4.880*	75.76(3.49)	74.16(3.70)	74.96(3.68)
2.	Eye height	2.931 ns	65.06(3.93)	63.74(3.73)	64.40(3.89)
3.	Shoulder height	2.947 ns	51.52(3.77)	50.26(3.53)	50.89(3.70)
4.	Elbow height**	31.862**	17.60(1.77)	20.32(2.88)	18.96(2.75)
5.	Thigh height	3.016 ns	11.51(1.35)	12.10(1.96)	11.80(1.71)
6.	Knee height	2.164 ns	46.30(4.27)	47.66(4.89)	46.98(4.64)
7.	Poplietal height	0.129 ns	41.42(4.21)	41.14(3.51)	41.28(3.87)
8.	Buttock-leg length	2.011 ns	94.74(5.74)	96.40(5.89)	95.57(5.88)
9.	Buttock-knee length	0.642 ns	49.22(3.79)	49.88(4.37)	49.55(4.11)
10.	Buttock-poplietal length	0.827 ns	43.64(3.66)	42.92(4.18)	43.28(3.95)

Note: ** = Significant at 1 per cent level of significance,
* = Significant at 5 per cent level of significance
Values in parentheses indicate the standard deviation of Mean

Table 4.6a : Percentile values of sitting anthropometric measurements of Uttarakhand women

S. No	Sitting anthropometric measurements (cm)	Percentile								
		Garhwal			Kumaon			Uttarakhand		
		5th	50th	95th	5th	50th	95th	5th	50th	95th
1.	Sitting height*	70.4	76.0	80.0	67.0	75.0	79.5	68.9	75.0	80.0
2.	Eye height	58.4	65.0	71.1	57.0	64.0	69.1	57.0	65.0	70.1
3.	Shoulder height	44.4	52.0	57.0	44.4	51.0	54.0	44.0	51.0	57.0
4.	Elbow height **	15.0	18.0	20.0	16.0	20.0	25.0	15.9	18.0	25.0
5.	Thigh height	9.5	11.0	14.0	8.0	12.0	15.0	8.9	12.0	15.0
6.	Knee height	40.9	45.0	53.6	41.0	48.0	55.1	40.9	46.0	55.0
7.	Poplietal height	36.4	40.5	49.6	36.0	41.0	46.5	36.0	41.0	47.0
8.	Buttock-leg length	88.0	93.0	105.0	87.9	97.0	105.0	87.9	95.0	106.0
9.	Buttock-knee length	44.4	49.0	57.1	43.0	50.5	56.0	43.0	50.0	56.1
10.	Buttock-poplietal length	38.9	43.0	50.6	35.8	42.0	50.0	37.9	42.5	50.1

Note: ** = Significant at 1 per cent level of significance,
* = Significant at 5 per cent level of significance
Values in parentheses indicate the standard deviation of Mean.

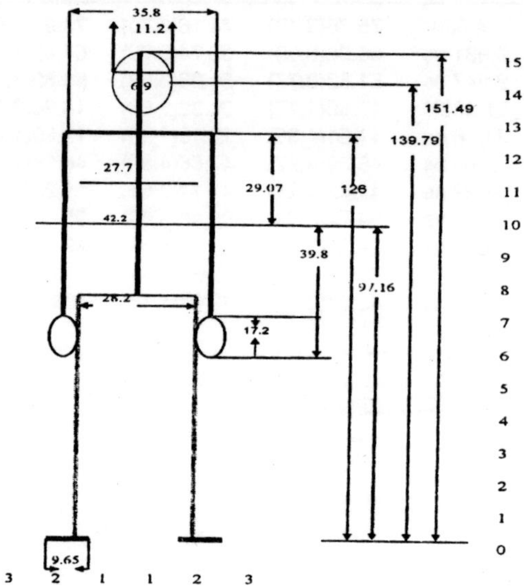

Fig. 4.4 : The line diagram illustration of Uttarakhand women (20-45yrs) in standing posture based on their mean anthropometric measurements

Note : All the measurements are in centimeters. Scale: 1:10

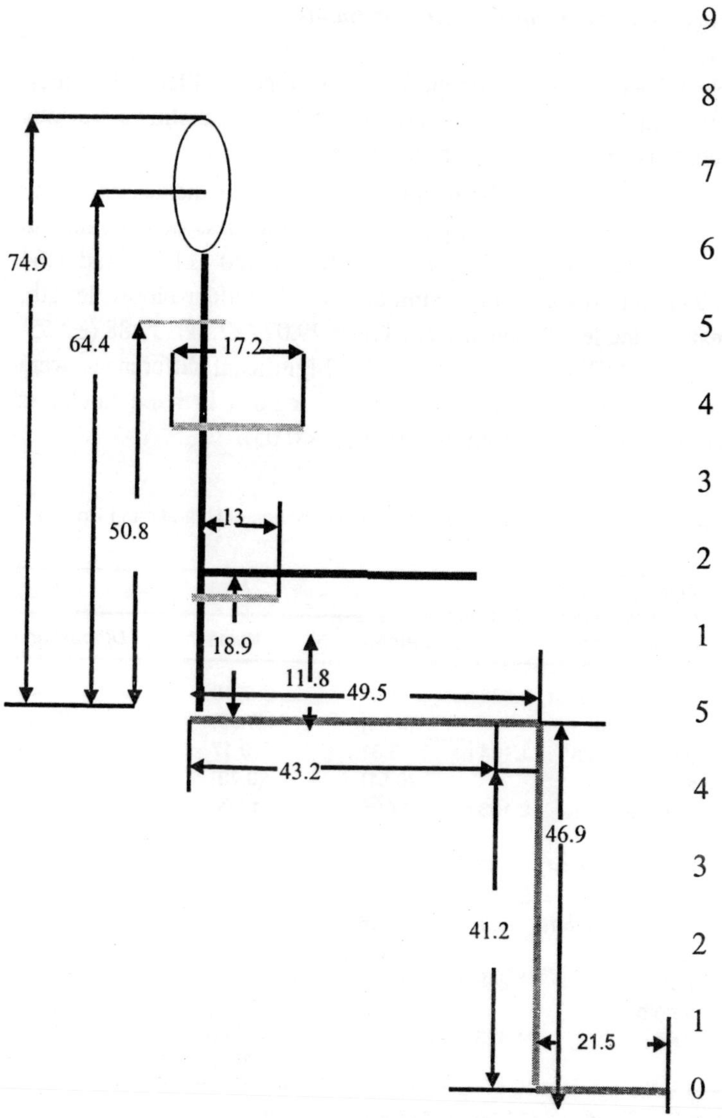

Fig. 4.5 : The line diagram illustration of Uttarakhand women (20-45 yrs) in sitting posture based on their mean anthropometric measurements

Note: All the measurements are in centimeters. Scale: 1:10

Anthropometric measurements of hands

Table 4.7 shows the anthropometric measurements of hand (F-values, mean and standard deviation) and Table 4.7a shows various percentiles values (5th, 50th and 95th).

It was found that the mean hand breadth at metacarpal, hand breadth at thumb, hand length and hand thickness of Uttarakhand women were 7.49 (±0.53), 9.24 (±0.64), 17.20 (±1.00) and 1.64 (±0.21) cm, respectively. Similarly the shoulder-elbow length, forearm-hand length and arm reach was 29.07 (±2.31), 39.88 (±2.55) and 74.38 (±3.53) cm, respectively. Significant differences were found between Garhwal and Kumaon regions in hand thickness ($p<0.01$) and shoulder-elbow length ($p<0.05$).

Table 4.7: Anthropometric measurements of hands of Uttarakhand women

S. No.	*Anthropometric measurements of hands (cm)*	*F- value*	*Mean*		
			Garhwal	*Kumaon*	*Uttarakhand*
1.	Hand breadth at metacarpal	0.036 ns	7.50 (0.36)	7.48 (0.65)	7.49 (0.53)
2.	Hand breadth at thumb	1.194 ns	9.31 (0.51)	9.17 (0.73)	9.24 (0.64)
3.	Hand length	2.103 ns	17.35 (0.74)	17.06 (1.19)	17.20 (1.00)
4.	Hand thickness**	31.831**	1.533 (0.21)	1.748 (0.16)	1.64 (0.21)
5.	Shoulder-elbow length*	6.881*	29.66 (2.30)	28.48 2.17	29.07 (2.31)
6.	Forearm-hand length	1.242 ns	40.17 (2.32)	39.6 (2.74)	39.88 (2.55)
7.	Arm reach	1.423 ns	73.96 (3.87)	74.8 (3.10)	74.38 (3.53)

Note: ** = Significant at 1 per cent level of significance,
* = Significant at 5 per cent level of significance
Values in parentheses indicate the standard deviation of Mean

Table 4.7a : Percentile values of anthropometric measurements of hands of Uttarakhand women

S. No.	*Anthropometric measurements of hands (cm)*	*Percentile*								
		Garhwal			*Kumaon*			*Uttarakhand*		
		5th	50th	95th	5th	50th	95th	5th	50th	95th
1.	Hand breadth at metacarpal	7.0	7.5	8.0	6.5	7.5	8.5	6.5	7.5	8.5
2.	Hand breadth at thumb	8.5	9.0	10.0	8.0	9.0	10.0	8.0	9.0	10
3.	Hand length	16.0	17.5	18.2	15.45	17.0	19.0	16	17.0	19.0
4.	Hand thickness**	1.2	1.5	1.9	1.5	1.7	2.0	1.25	1.7	2.0
5.	Shoulder -elbow length*	25.0	30	33.1	25.0	29.0	32.0	25	29.0	32.0
6.	Forearm-hand length	36.9	40.0	44.0	35.0	40.0	43.0	35.0	40.0	44.0
7.	Arm reach	68.9	74.0	80.0	70.4	75.0	79.5	69.9	74.5	80.0

Note: ** = Significant at 1 per cent level of significance,
* = Significant at 5 per cent level of significance
Values in parentheses indicate the standard deviation of Mean

Body breadths

Table 4.8a shows the body breadths (F-values, mean and standard deviation) and Table 4.8b shows various percentiles values (5th, 50th and 95th).

It was found that the interpupillary distance, head length and head breadth were 6.90 (±1.03), 19.88 (±1.73) and 11.28 (±2.10) cm, respectively. Similarly, the shoulder breadth, chest breadth, maximum body breadth, elbow-elbow breadth, sitting hip breadth, standing hip breadth and knee-knee breadth were 35.86 (±2.81), 27.72 (±2.30), 42.25 (±4.10), 38.10 (±4.67), 32.31 (±2.47), 28.21 (±3.06) and 13.29 (±1.96) cm, respectively. Significant differences were found between Garhwal and Kumaon regions in head breadth, shoulder breadth, elbow-elbow breadth, knee-knee breadth and sitting hip breadth ($p<0.01$) showing that women from Kumaon region have broader body size than Garhwal region.

Table 4.8 : Body breadths of Uttarakhand women

S. No.	Body breadths (cm)	F-value	Mean (S.D)		
			Garhwal	Kumaon	Uttarakhand
1.	Interpupillary distance	0.414 ns	6.97 (1.41)	6.84 (0.41)	6.90 (1.03)
2.	Head length	1.082 ns	20.06 (1.76)	19.70 (1.67)	19.88 (1.73)
3.	Head breadth**	23.463 **	10.36 (2.01)	12.20 (1.75)	11.28 (2.10)
4.	Shoulder breadth**	30.156 **	34.5 (2.26)	37.22 (2.64)	35.86 (2.81)
5.	Chest breadth	3.086 ns	28.12 (1.76)	27.32 (2.66)	27.72 (2.30)
6.	Maximum body breadth	1.710 ns	41.72 (3.89)	42.79 (4.23)	42.25 (4.10)
7.	Elbow-elbow breadth**	29.461**	35.87 (3.89)	40.34 (4.28)	38.10 (4.67)
8.	Hip breadth sitting**	14.697**	31.42 (1.73)	33.20 (2.75)	32.31 (2.47)
9.	Hip breadth standing	2.169 ns	27.92 (1.33)	28.50 (2.41)	28.21 (3.06)
10.	Knee-Knee breadth**	41.850**	12.22 (1.09)	14.36 (2.04)	13.29 (1.96)

Note: ** = Significant at 1 per cent level of significance,
Values in parentheses indicate the standard deviation of mean

Table 4.8a: Percentile values of body breadths of Uttarakhand women

S. No.	Body breadths (cm)	Percentile								
		Garhwal			Kumaon			Uttarakhand		
		5th	50th	95th	5th	50th	95th	5th	50th	95th
1.	Interpupillary distance	6.5	7.0	7.2	6.0	7.0	7.5	6.0	7.0	7.5
2.	Head length	18.5	20.0	22.0	17.0	20.0	22.0	17.0	20.0	22.0
3.	Head breadth**	8.7	10.0	12.0	10.0	12.0	15.0	9.0	11.0	15.0
4.	Shoulder breadth**	32.0	34.0	39.0	33.0	38.0	41.5	32.0	35.0	41.0
5.	Chest breadth	25.0	28.0	31.0	24.0	27.5	31.5	24.0	28.0	31.0
6.	Maximum body breadth	34.5	43.0	47.5	36.9	43.0	48.0	35.0	43.0	48.0
7.	Elbow-elbow breadth**	29.0	36.0	42.0	32.0	40.0	48.5	30.9	39.0	45.1
8.	Sitting Hip breadth**	28.5	32.0	33.0	29.0	33.0	37.5	28.9	32.0	37.0
9.	Standing Hip breadth	25.5	28.0	30.0	24.4	28.5	32.0	25.0	28.0	35.1
10.	Knee-knee breadth**	11.0	12.0	14.0	12.0	14.0	18.0	11.0	13.0	17.0

Note: ** = Significant at 1 per cent level of significance,
Values in parentheses indicate the standard deviation of mean

Body circumferences and depths

Table 4.9 shows the various body circumferences and depths (F-values, mean and standard deviation) and Table 4.9a shows various percentiles values (5th, 50th and 95th).

It was found that the chest and waist circumferences were 82.10 (±6.50) and 70.59 (±7.44) cm whereas various body depths such as chest depth, waist depth, maximum body depth and buttock depth were 18.00 (±3.06), 13.63 (±2.47), 22.28 (±3.36) and 13.75 (±2.22) cm, respectively. Foot length and breadth was 21.54 (±1.53) and 9.65 (±0.98) cm, respectively. Significant differences were found between Garhwal and Kumaon region in waist circumference, chest depth, waist depth, maximum body depth and foot breadth ($p<0.01$) besides waist depth ($p<0.05$).

Table 4.9: Body circumferences and depths of Uttarakhand women

S. No.	*Body circumferences and depths (cm)*	*F-value*	*Mean (S.D)*		
			Garhwal	*Kumaon*	*Uttarakhand*
1.	Chest circumference	2.418 ns	81.10 (5.43)	83.11 (7.27)	82.10 (6.50)
2.	Waist circumference**	19.095 **	67.60 (6.65)	73.58 (6.95)	70.59 (7.44)
3.	Chest depth**	11.292 **	17.02 (3.13)	18.98 (2.64)	18.00 (3.06)
4.	Waist depth*	6.396 *	13.02 (2.23)	14.24 (2.54)	13.63 (2.47)
5.	Maximum body depth**	26.203 **	23.82 (2.66)	20.74 (3.28)	22.28 (3.36)
6.	Buttock depth	0.181 ns	13.64 (1.94)	13.80 (1.8)	13.75 (2.22)
7.	Foot breadth**	13.537 **	9.99 (0.87)	9.31 (0.95)	9.65 (0.98)
8.	Foot length	0.016 ns	21.56 (1.73)	21.52 (1.29)	21.54 (1.53)

Note: ** = Significant at 1 per cent level of significance,
* = Significant at 5 per cent level of significance
Values in parentheses indicate the standard deviation of mean

Table 4.9a : Percentile values of body circumferences and depths of Uttarakhand women

S. No.	Body circumferences and depths (cm)	Garhwal 5th	Garhwal 50th	Garhwal 95th	Kumaon 5th	Kumaon 50th	Kumaon 95th	Uttarakhand 5th	Uttarakhand 50th	Uttarakhand 95th
1.	Chest circumference	73.5	80.5	90.0	70.9	84.0	94.5	71.9	82.0	94.1
2.	Waist circumference**	58.0	67.5	77.5	62.9	73.0	86.2	60.0	70.0	84.0
3.	Chest depth**	12.5	17.0	21.5	15.0	19.0	23.0	13.0	18.0	23.0
4.	Waist depth*	10.0	12.0	17.1	11.0	14.0	19.0	10.0	13.0	18.0
5.	Maximum body depth**	19.5	24.0	28.0	16.0	21.0	25.5	16.0	23.0	28.0
6.	Buttock depth	11.0	13.0	16.0	12.0	14.0	17.0	11.0	13.0	17.0
7.	Foot breadth**	8.5	10.0	11.0	7.4	9.5	10.5	8.0	10.0	11.0
8.	Foot length	19.0	22.0	24.5	19.4	22.0	23.5	19.0	22.0	24.0

Note: ** = Significant at 1 per cent level of significance, *= Significant at 5 per cent level of significance Values in parentheses indicate the standard deviation of mean

It could, therefore, be inferred that though the significant differences were not observed in most of the standing and sitting heights yet a few body depths and most of body breadth which showed broader body type of women from Kumaon region probably due to the difference in the geographical conditions and comparatively less physical activity of Kumaon women than Garhwal women because of poor infrastructure facilities in Garhwal region.

6. Comparison of Anthropometric Measurements of Uttarakhand Women vis-á-vis Women of other States

Tables 4.10 and 4.10a revealed comparison of standing and sitting anthropometric measurements, respectively of hill women of Uttarakhand with women of other states.

A total of 16 anthropometric measurements were selected for comparative analysis which included 5 standing anthropometric measurements, 3 hand measurements and 8 sitting anthropometric measurements. All the anthropometric measurements of other states were database of different centers of AICRP, Home Science. All the sitting heights measurements of other states were taken from the floor but in this particular study the sitting height measurements

Table 4.10: Comparison of standing anthropometric measurements of hill women of Uttarakhand with women of other states

S. No.	Standing anthropometric measurements (cm)	Mean						
		*Uttarakhand**	*Haryana (Hisar)*	*Punjab (Ludhiana)*	*Andhra Pradesh (Hyderabad)*	*Rajasthan (Udaipur)*	*Uttarakhand** (Pantnagar)*	*Karnataka (Dharwad)*
		N=100	*N=1000*	*N=1000*	*N=1000*	*N=1000*	*N=1000*	*N=1000*
1.	Weight (kg)	40.27	46.72+6.45	53.30+13.03	43.66+3.39	43.72+3.45	46.32+6.05	43.90+3.63
2.	Height	151.49	157.44+5.95	156.79+5.3	152.54+1.05	151.99+0.5	151.97+0.48	149.40-2.09
3.	Eye height	139.79	145.91+6.12	144.08+4.29	142.46+2.67	141.78+1.99	140.21+0.42	139.29-0.5
4.	Shoulder height	126.00	132.50+6.5	130.81+4.81	127.61+1.61	126.99+0.99	126.26+0.26	126.74+0.74
5.	Standing elbow height	97.16	112.89+15.73	99.69+2.59	99.44+2.28	98.90+1.74	95.75-1.41	97.21+0.05
6.	Hand length	17.35	18.07+0.72	15.06-2.29	17.60+0.25	18.55+1.2	18.56+1.21	18.11+0.76
7.	Forearm length	40.17	44.94+4.77	39.52-0.65	40.98+0.81	39.92-0.25	39.17-1.0	41.13+0.96
8.	Arm length	73.96	71.56-2.4	67.25-6.71	68.19-5.77	67.44-6.52	66.80-7.16	68.69-5.27

Note: All the anthropometric measurements of other states are database of AICRP, Home Science. Values in bold letters indicate difference between anthropometric measurements of Uttarakhand and other states.* indicates data for Uttarakhand (hill women) collected in the present study. ** indicates data for Uttarakhand (Pantnagar centre) for *Tarai* region.

Table 4.10a : Comparison of sitting anthropometric measurements of hill women of Uttarakhand with women of other states

S. No.	Sitting anthropometric measurements (cm)	Mean						
		*Uttarakhand**	*Haryana (Hisar)*	*Punjab (Ludhiana)*	*Andhra Pradesh (Hyderabad)*	*Rajasthan (Udaipur)*	*Uttarakhand** (Pantnagar)*	*Karnataka (Dharwad)*
		N=100	*N=1000*	*N=1000*	*N=1000*	*N=1000*	*N=1000*	*N=1000*
1.	Sitting height	116.76	118.89+2.13	118.77+2.01	117.25+0.49	116.89+0.13	117.53+0.77	114.93-1.83
2.	Shoulder height	92.52	115.27+22.75	93.56+1.04	92.67+0.15	94.13+1.61	92.31-0.21	93.12+0.6
3.	Eye height	106.06	116.13+10.07	105.45-0.61	106.91+0.85	107.80+1.74	105.84-0.22	104.19-1.87
4.	Elbow height	57.6	60.70+3.1	62.55+4.95	63.49+5.89	65.42+7.82	62.10+4.5	64.03-6.43
5.	Knee height	46.3	50.41+4.11	48.35+2.05	49.24+2.94	48.26+1.96	46.71+0.41	45.54-0.76
6.	Thigh height	11.51	12.43+0.92	8.61-2.9	13.23+1.72	14.62+3.11	11.50-0.01	11.84+0.33
7.	Buttock-knee length	49.22	56.24+7.02	50.70+1.48	53.50+4.28	50.31+1.09	50.45+1.23	49.05-0.17
8.	Buttock-poplietal length	43.64	48.54+4.9	41.48-2.16	47.34+3.7	45.13+1.49	42.54-1.1	43.30-0.34

Note: All the anthropometric measurements of other states are database of AICRP, Home Science. Values in bold letters indicate difference between anthropometric measurements of Uttarakhand and other states.* indicates data for Uttarakhand (hill women) collected in the present study. ** indicates data for Uttarakhand (Pantnagar centre) for *Tarai* region.

were taken from the seat surface, therefore for comparison 41cm (seat height of chair from floor/ poplietal height sitting) was added as a constant and added to the sitting height measurements of hill women of Uttarakhand. Values in bold letters indicate difference between anthropometric measurements of hill women of Uttarakhand and other states. Positive (+) values indicate that the particular anthropometric measurements of that state is greater than the anthropometric measurement of hill women of Uttarakhand where-as negative (-) values indicate that the particular anthropometric measurement of that state is lower than the anthropometric measurement of hill women of Uttarakhand.

It was found that all the anthropometric measurements of Haryana (Hissar) women were greater than the anthropometric measurements of hill women of Uttarakhand except arm length.

It was also found that besides sitting eye height, sitting thigh height, buttock-popelietal length and 3 hand measurements (hand, forearm and total arm length) of Punjab (Ludhiana) women, all other anthropometric measurements were greater than hill women of Uttarakhand. Also all the anthropometric measurements of Andhra Pradesh (Hyderabad) women were greater than hill women of Uttarakhand except arm length.

Besides arm length and forearm length all other anthropometric measurements of Rajathan (Udaipur) women were greater than hill women of Uttarakhand. Also besides standing elbow height, sitting shoulder height , sitting eye height, sitting thigh height, buttock-poplietal length, arm length and forearm length all other anthropometric measurements of Uttarakhand (Pantnagar, *Tarai* region) women were greater than hill women of Uttarakhand.

Also, except standing height, standing eye height, sitting height, sitting eye height, sitting elbow height, sitting knee height, buttock-knee length, buttock-poplietal length and arm length all other anthropometric measurements of Karnataka (Dharwad) women were greater than hill women of Uttarakhand.

It could be concluded that all the selected standing anthropometric measurements of other states were higher than those of Uttarakhand women except standing height and eye height of Karnataka (Dharwad) women. Also all the selected sitting anthropometric measurements of Andhra Pradesh (Hyderabad), Rajasthan (Udaipur) and Haryana (Hissar) were higher than the hill women of Uttarakhand.

7. Demographic Profile of Subjects Employed in Laboratory Experiments

Table 4.11 showed the demographic profile of the women subjects employed for laboratory experiments. It was found that the mean age of subjects was 35±3 years. All the subjects were illiterate and belonged to nuclear families having medium family size (4-7 members) except one who had a large family size (>8 members).The main source of income was service, agriculture and livestock. The average mean time spent for collecting fuel and fodder was 4.5 h/day whereas it was 3.5 h/day for agriculture activities.

Table 4.11: Demographic profile of the subjects employed for laboratory experiments

S. No.	*Demographic characteristics*	*Frequency*
1.	**Mean Age (years)**	35±3 years
2.	**Education**	
i	Illiterate	7
ii	Primary	
iii	Junior high school	
iv	High school	
v	Intermediate	
vi	Graduate	
vii	Post-graduate	
3.	**Family Type**	
i	Nuclear	7
ii	Joint	
iii	Extended	
4.	**Family Size**	
i	1-3 (Small)	
ii	4-7 (Medium)	6
iii	> 8 (Large)	1
5.	**Source of Income**	
i	Agriculture	3
ii	Business	-
iii	Service	6
iv	Livestock (dairy)	7
6.	**Monthly Income (Rs)**	
i	500-1000	
ii	1001-2000	
iii	2001-3000	2
iv	3001-4000	3
v	4001-5000	1
vi	>5000	1
7.	**Average Time Spent (h/day)**	
i	Collecting fuel, fodder etc.	4.5
ii	Agricultural activities	3.5

8. Anthropometric Measurements of Subjects Employed in Laboratory Experiments

Table 4.12 revealed anthropometric measurements of the women subjects employed for laboratory experiments. The mean weight, height, BMI, BSA were found as 51.42 kg, 156.71 cm, 20.89 kg/ m^2 and 1.49 m^2, respectively.

Table 4.12: Anthropometric measurement of the subjects employed for laboratory experiments

S. No.	*Anthropometric measurements*							
	Weight (kg)	*BMI (kg/m²)*	*BSA (m²)*	*Height (cm)*	*Eye height (cm)*	*Shoulder height (cm)*	*Standing elbow height (cm)*	*Standing knuckle height (cm)*
1.	57	22.55	1.58	159	147	133	108	75
2.	51	18.96	1.54	164	152	137	112	78
3.	57	22.83	1.57	158	145	131	104	72
4.	43	18.86	1.35	151	139	128	96	69
5.	45	19.74	1.38	151	138	127	97	70
6.	58	21.83	1.62	163	150	133	108	78
7.	49	21.49	1.43	151	138	126	99	72
Mean	51.42	20.89	1.49	156.71	144.14	130.71	103.42	73.42

9. Environmental Conditions during Laboratory Experiments

Table 4.13 depicts the environmental conditions during the experiment. The temperature and humidity ranged from 30.7-33.8 ºC and 43-51%, respectively.

Table 4.13 : Existing environmental conditions during the experiment

S. No.	*Environmental Parameters*	*Range*	
		Minimum	*Maximum*
1.	Temperature (ºC)	30.7	33.8
2.	Humidity (%)	43	51
3.	Light (Lux)	105	200
4.	Sound (dB)	42.5	52.3

10. Effect of Interaction of Load, Mode and Grade on Physiological Responses and Perceived Exertion

Tables 4.14 to 4.19 show the effect of interaction of three main factors i.e. load, mode and grade on physiological responses and perceived exertion. A total of 27 different trials were conducted to observe the variation due to interaction of loads, modes and grades on five physiological parameters i.e. Heart Rate (HR), Oxygen Consumption Rate (OCR), Energy Expenditure Rate (EER), OCR per kg body weight, Metabolic Equivalents (METS) as well as on Rating of Perceived Exertion (RPE).

Heart rate and increase in heart rate from resting level

The results of experiments conducted on various modes of carrying load with different loads on different grades were analyzed and presented in Table 4.14 for mean heart rate and increase in mean heart rate from its resting level. Fig. 4.6 showed the mean heart rate, respectively in various modes of carrying loads on different grades. The analysis of variance is shown in Appendix B, Tables B-1 and B-2.

Exerting human energy in any form causes an increase in heart rate. This response was selected because it could be monitored without stopping the activity and disturbing the subject. The heart rate was measured with the help of polar heart rate monitor from 5th minute till 8th minute at each minute and then the mean of 5th, 6th, 7th and 8th minute heart rate was taken for each subject. Again the subject's mean was averaged for each test. The mean resting heart rate was noted as 77 beats/min. The variation of this response at different modes and grades of carrying loads are described below:

(a) Carrying loads on back

- On H-grade (horizontal surface) the mean heart rate (range) was 106 (100-110), 112 (103-122) and 116 (108-127) beats/min whereas mean increase in heart rate was 28 (25-33), 35 (28-45), and 39 (32-49) beats/min with 15, 20 and 25

Table 4.14: Physiological response in terms of heart rate during various modes of carrying loads

Modes of carrying load	*Heart rate (beats/min) Load (kg)*					
	15		*20*		*25*	
	Mean HR and Range	*Increase in Mean HR and Range*	*Mean HR and Range*	*Increase in Mean HR and Range*	*Mean HR and Range*	*Increase in Mean HR and Range*
Back						
H	106 100-110	28 25-33	112 103-122	35 28-45	116 108-127	39 32-49
G_1	112 103-119	35 27-42	118 112-132	41 35-54	123 115-131	45 39-54
G_2	122 109-127	44 36-48	130 117-143	53 41-65	135 121-148	58 44-70
Head						
H	109 103-115	32 27-38	116 106-126	39 31-49	121 115-131	43 39-53
G_1	116 107-124	39 30-47	123 115-137	46 39-59	128 121-141	51 45-63
G_2	125 113-131	47 36-56	135 122-150	58 46-73	140 128-153	62 51-75
Shoulder						
H	115 110-124	37 31-47	124 117-134	47 39-57	127 126-138	50 44-60
G_1	121 114-135	44 38-58	130 124-144	53 46-66	138 131-152	61 54-74
G_2	134 120-147	57 44-70	145 135-158	68 59-80	150 140-161	72 64-83

Resting mean heart rate=77 beats/min, No. of subjects=7
H = Zero grade (Horizontal surface), $\mathbf{G_1}$ = 5% Grade, $\mathbf{G_2}$ = 10% grade

kg of loads, respectively. The range of mean and increase in heart rate were 100-127 and 25-49 beats/min, respectively.

- On 5 per cent grade (G_1) the mean heart rate was 112 (103-119), 118 (112-132) and 123 (115-131) beats/min whereas mean increase in heart rate was 35 (27-42), 41 (35-54) and 45 (39-54) beats/min with 15, 20 and 25 kg of loads, respectively. The range of mean and increase in heart rate was 103-131 and 27-54 beats/min, respectively.

- On 10 per cent grade (G_2) the mean heart rate was 122 (109-127), 130 (117-143) and 135 (121-148) beats/min where as mean increase in heart rate was 44 (36-48), 53 (41-65) and 58 (44-70) beats/min with 15, 20 and 25 kg of loads, respectively. The range of mean and increase in heart rate was 109-148 and 36-70 beats/min, respectively.

It is observed that during carrying loads on the back, the highest mean and increase in heart rate was found with 25 kg load on 10% grade and the lowest with 15 kg load on zero grade (horizontal surface).

(b) Carrying loads on head

- On H-grade (horizontal surface) the mean heart rate was 109 (103-115), 116 (106-126) and 121 (115-131) beats/min whereas the mean increase in heart rate was 32 (27-38), 39 (31-49), and 43 (39-53) beats/min with 15, 20 and 25 kg of loads, respectively. The range of mean and increase in heart rate was 103-131 and 27-53 beats/min, respectively.
- On 5 per cent grade (G_1) the mean heart rate was 116 (107-124), 123 (115-137), and 128 (121-141) beats/min whereas the mean increase in heart rate was 39 (30-47), 46 (39-59) and 51 (45-63) beats/min with 15, 20 and 25 kg of loads, respectively. The range of mean and increase in heart rate was 107-141 and 30-63 beats/min, respectively.
- On 10 per cent grade (G_2) the mean heart rate was 125 (113-131), 135 (122-150) and 140 (128-153) beats/min whereas mean increase in heart rate was 47 (36-56), 58 (46-73) and 62 (51-75) beats/min with 15, 20 and 25 kg of loads, respectively. The range of mean and increase in heart rate was 113-153 and 36-75 beats/min, respectively.

So, it was found that during carrying loads on head also the highest mean and increase in heart rate was found with 25 kg load on 10 per cent grade and the lowest with 15 kg load on H-grade.

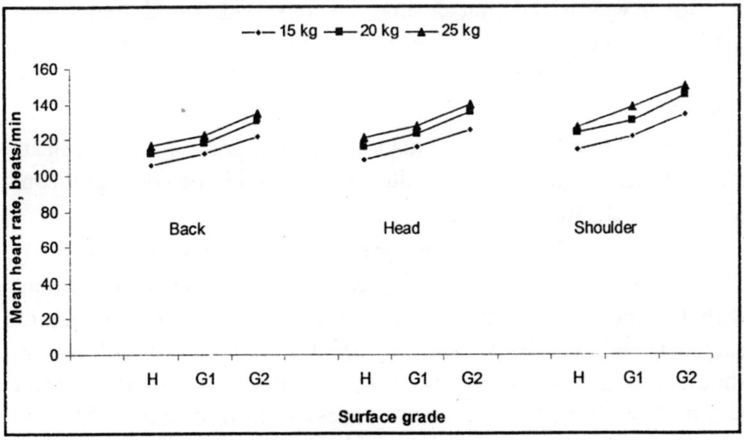

Fig. 4.6 : Variation in mean heart rate for different modes of carrying loads at various surface grades

(c) Carrying loads on shoulder

- On H-grade (horizontal surface) the mean heart rate was 115 (110-124), 124 (117-134) and 127 (126-138) beats/min whereas mean increase in heart rate was 37 (31-47), 47 (39-57) and 50 (44-60) beats/min with 15, 20 and 25 kg of loads, respectively. The range of mean and increase in heart rate was 110-138 and 31-60 beats/min, respectively.
- On 5 per cent grade (G_1) the mean heart rate was 121 (114-135), 130 (124-144) and 138 (131-158) beats/min whereas mean increase in heart rate was 44 (38-58), 53 (46-66) and 61 (54-74) beats/min with 15, 20 and 25 kg of loads, respectively. The range of mean and increase in heart rate was 114-152 and 38-74 beats/min, respectively.
- On 10 per cent grade (G_2) the mean heart rate was 134 (120-147), 145 (135-158) and 150 (140-161) beats/min whereas mean increase in heart rate was 57 (44-70), 68

(59-80) and 72 (64-83) beats/min with 15, 20 and 25 kg of loads, respectively. The range of mean and increase in heart rate was 120-161 and 44-83 beats/min, respectively.

So, it was found that during carrying loads on shoulder also the highest mean and increase in heart rate was found with 25 kg load on 10 per cent grade and the lowest was found with 15 kg load on H-grade.

The interaction effect of load, mode and grade for mean heart rate and increase in heart rate values were found to be non-significant (Appendix B, Table B-1).

It could be concluded that the mean and increase in mean heart rate increase linearly with the increase in load, this is in agreement with Borghols *et al,* (1978) and grade in all the three modes of carrying loads viz. back, head and shoulder. The carrying loads on shoulder was physiologically most demanding followed by head and back with all the loads and grades. Therefore, the back mode of carrying loads was physiologically least demanding, this is in agreement with Singh and Kaul (1972).

Oxygen consumption rate

Table 4.15 shows the variation in OCR (mean and range) and increase in mean OCR from its resting level and range during various modes of carrying loads on different grade and depicted in Fig. 4.7. Analysis of variance is given in Appendix B, Table B-3. The mean resting OCR was 0.219 l/min.

(a) Carrying loads on back

- On H-grade (horizontal surface) the mean OCR and range was 0.62 (0.58-0.64), 0.71 (0.64-0.76), and 0.77 (0.69-0.80) l/min, and increase in OCR was 0.41 (0.36-0.42), 0.50 (0.42-0.54) and 0.55 (0.47-0.58) l/min with 15, 20 and 25 kg loads, respectively. In general, the OCR value and increase in value of OCR ranged from 0.58-0.80 and 0.36-0.58 l/min, respectively while carrying loads on back on H-grade.

Table 4.15 : Physiological response in terms of oxygen consumption rate during various modes of carrying loads

Modes of carrying loads	*Oxygen consumption rate (min) Load (kg)*					
	15		*20*		*25*	
	Mean OCR and Range	*Increase in OCR and Range*	*Mean OCR and Range*	*Increase in OCR and Range*	*Mean OCR and Range*	*Increase in OCR and Range*
Back						
H	0.62 0.58 - 0.64	0.41 0.36-0.42	0.71 0.64-0.76	0.50 0.42-0.54	0.77 0.69-0.80	0.55 0.47-0.58
G_1	0.72 0.65-0.85	0.50 0.43-0.63	0.80 0.76-0.87	0.58 0.54-0.65	0.97 0.74-0.99	0.75 0.52-0.72
G_2	0.89 0.77-0.96	0.67 0.55-0.74	0.94 0.84-1.02	0.72 0.62-0.8	1.07 1.04-1.15	0.85 0.82-0.93
Head						
H	0.67 0.63-0.69	0.45 0.41-0.47	0.80 0.68-0.88	0.58 0.46-0.66	0.84 0.74-0.89	0.62 0.52-0.67
G_1	0.84 0.71-0.89	0.62 0.49-0.67	0.93 0.89-0.96	0.71 0.67-0.74	1.03 0.95-1.12	0.81 0.73-0.90
G_2	0.99 0.81-1.08	0.77 0.59-0.86	1.10 0.91-1.27	0.88 0.69-0.86	1.26 1.16-1.32	1.04 0.94-1.10
Shoulder						
H	0.82 0.69-0.94	0.60 0.47-0.72	0.89 0.74-0.98	0.67 0.52-0.76	1.00 0.95-1.03	0.78 0.73-0.81
G_1	0.92 0.74-0.99	0.70 0.52-0.77	1.05 1.00-1.09	0.83 0.78-0.87	1.16 1.10-1.24	0.94 0.88-1.02
G_2	1.09 0.90-1.18	0.77 0.68-0.96	1.29 1.16-1.42	1.07 0.94-1.28	1.42 1.27-1.50	1.20 1.05-1.28

Note: As subjects hold the handrails while walking on treadmill it causes an underestimation of VO_2 by 15 %
Resting OCR at resting level = 0.219 l/min, No. of subjects = 7
H = Zero grade (Horizontal surface), $\mathbf{G_1}$ = 5% Grade, $\mathbf{G_2}$ = 10% grade

- On 5 per cent grade (G_1) the mean OCR and range was 0.72 (0.65-0.85), 0.80 (0.76-0.87) and 0.97 (0.74-0.99) l/min and increase in OCR was 0.50 (0.43-0.63), 0.58 (0.54-

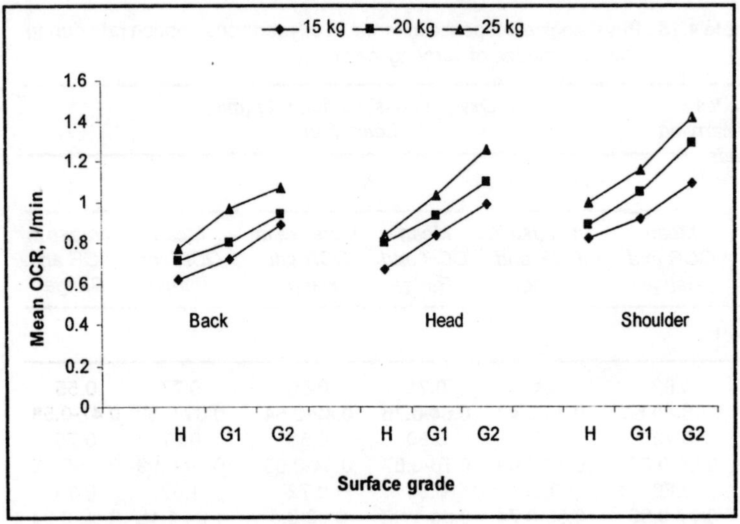

Fig. 4.7 : Variation in mean oxygen consumption rate for different modes of carrying loads on various surface grades

0.65) and 0.75 (0.52-0.72) l/min with 15, 20 and 25 kg loads, respectively. In general, the OCR value and increase in value of OCR ranged from 0.65-0.94 and 0.43-0.72 l/min, respectively while carrying loads on back on G_1-grade.

➢ On 10 pér ent grade (G_2) the mean OCR and range was 0.89 (0.77-0.96), 0.94 (0.84-1.02) and 1.07 (1.04-1.15) l/min and increase in OCR was 0.67 (0.55-0.74), 0.72 (0.62-0.80) and 0.85 (0.82-0.93) l/min with 15, 20 and 25 kg loads, respectively. In general, the OCR value and increase in value of OCR ranged from 0.77-1.15 and 0.55-0.93 l/min, respectively while carrying loads on back on G_2-grade.

It is, therefore, concluded that with the increase in load and grade the mean value of OCR also increased while carrying load on the back. The highest mean value of OCR was found with 25 kg load on G_2-grade whereas the lowest was with 15 kg load for horizontal grade. The value of OCR during carrying loads on head on different surface grades ranged from 0.58-1.15 and increase in the value of OCR from its resting level ranged from 0.36-0.85 l/min.

(b) Carrying loads on head

- On H-grade (horizontal surface) the mean OCR and range was 0.67 (0.58-0.64), 0.80 (0.68-0.88) and 0.84 (0.74-0.89) l/min, and increase in OCR was 0.45 (0.14-0.47), 0.58 (0.46-0.66) and 0.62 (0.52-0.67) l/min with 15, 20 and 25 kg loads, respectively. The OCR value and increase in value of OCR ranged from 0.63-0.89 and 0.41-0.67 l/min, respectively during carrying loads on back on H-grade.
- On 5 per cent grade (G_1) the mean (range) OCR was 0.84 (0.71-0.89), 0.93 (0.89-0.96) and 1.03 (0.95-1.12) l/min and increase in mean OCR (range) was 0.62 (0.49-0.67), 0.71 (0.67-0.74) and 0.81 (0.730.90) l/min with 15, 20 and 25 kg load, respectively. The OCR value and increase in value of OCR ranged from 0.71-1.12 and 0.49-0.90 l/min, respectively during carrying loads on back on G_1-grade.
- On 10 per cent grade (G_2) the mean OCR and range was 0.99 (0.81-1.08), 1.10 (0.91-1.27) and 1.26 (1.16-1.32) l/min and increase in OCR was 0.77 (0.59-0.86), 0.88 (0.69-0.86) and 1.04 (0.94-1.10) l/min with 15, 20 and 25 kg loads, respectively. The OCR value and increase in value of OCR ranged from 0.81-1.32 and 0.59-1.10 l/min, respectively during carrying loads on the back on G_2-grade.

It can be concluded that the OCR increases linearly as a function of the increasing load and grade while carrying loads on head as well as on the back. The highest mean value of 1.26 (1.16-1.32) l/min was found for carrying 25 kg load on G_2-grade whereas the lowest mean value 0.67 (0.63-0.69) l/min while carrying 15 kg load on H-grade.

(c) Carrying loads on shoulder

- On H-grade (horizontal surface) the mean OCR (range) was 0.82 (0.69-0.94), 0.89 (0.74-0.98), and 1.00 (0.95-1.03) l/min, and increase in OCR (range) was 0.60 (0.47-

0.72), 0.67 (0.52-0.76) and 0.78 (0.73-0.81) l/min with 15, 20 and 25 kg loads, respectively. The OCR value and increase in value of OCR ranged from 0.69-1.03 and 0.47-0.81 l/min, respectively during carrying loads on back.

- On 5 per cent grade (G_1) the mean (range) OCR was 0.92 (0.74-0.99), 1.05 (1.00-1.09) and 1.16 (1.10-1.24) l/min and increase in mean OCR (range) was 0.70 (0.52-0.77), 0.83 (0.78-0.87) and 0.94 (0.88-1.02) l/min with 15, 20 and 25 kg loads, respectively. The OCR value and increase in value of OCR ranged from 0.74-1.24 and 0.52-1.02 l/min, respectively during carrying loads on back on G_1 grade.
- On 10 per cent grade (G_2) the mean (range) OCR was 1.09 (0.90-1.18), 1.29 (1.16-1.42) and 1.42 (1.27-1.50) l/min and increase in mean (range) OCR was 0.77 (0.68-0.96), 1.07 (0.94-1.28) and 1.20 (1.05-1.28) l/min with 15, 20 and 25 kg loads, respectively. The OCR value and increase in value of OCR ranged from 0.90-1.50 and 0.68-1.28 l/min, respectively during carrying loads on back on G_2-grade. The interaction effect of mode, load and grade was found to be non-significant

It could be concluded that the highest mean value 1.42 (1.27-1.50) l/min was found for carrying 25 kg load on shoulder at G_2-grade which was more than two times demanding physiologically than the lowest mean value of 0.62 (0.58-0.64) l/min found for carrying 15 kg load on back at H-grade. Also, carrying loads on shoulder was more demanding in terms of OCR than the head and back modes on all the grades. Moreover, the 25 kg load on G_2-grade produced greatest metabolic load than other two loads and grades. Therefore, the back mode of carrying loads was physiologically (OCR) least demanding than the other two modes viz. head and shoulders.

Energy expenditure rate

Table 4.16 shows the mean Energy Expenditure Rate (EER) kcal/min and increase in EER from its resting level for various modes of carrying loads on different grades. Analysis of variance is shown in Appendix B, Table B-4. The mean EER at resting level was 1.107 kcal/min.

(a) Carrying loads on back

- On H-grade (horizontal grade) the mean EER (range) was 3.11 (2.97-3.25), 3.58 (3.27-3.72) and 3.90 (3.50-4.06) kcal/min and increase in EER (range) from resting level was 2.00 (1.87-2.15), 2.48 (2.17-2.62) and 2.80 (2.40-2.96) kcal/min with 15, 20 and 25 kg loads, respectively. The EER ranged from 2.97-4.06 kcal/min and increase from resting level ranged from 1.87-2.96 kcal/min.
- On G_1-grade the mean EER (range) was 3.65 (3.32-4.33), 4.04 (3.83-4.43) and 4.39 (3.75-4.74) kcal/min and increase in EER (range) from its resting level was 2.56 (2.22-3.23), 2.94 (2.73-3.33) and 3.29 (2.65-3.64) kcal/min with 15, 20 and 25 kg loads, respectively. The EER ranged from 3.32-4.74 kcal/min and increase from its resting level ranged from 2.22-3.64 kcal/min.
- On G_2-grade the mean EER (range) was 4.53 (3.91-4.86), 4.76 (4.26-5.17) and 5.44 (5.25-5.83) and increase in EER (range) from its resting level was 3.43 (2.81-3.76), 3.66 (3.16-4.07) and 4.34 (4.15-4.73) kcal/min with 15, 20 and 25 kg loads, respectively. The EER ranged from 3.91-5.83 kcal/min and the increase from resting level ranged from 2.81-4.73 kcal/min.

While considering EER values for carrying different loads on the back on different grades, it is evident that the highest value of 5.44 (5.25-5.83) kcal/min was found with 25 kg load on G_2-grade whereas the lowest value of 3.11 (2.97-3.25) kcal/min was found with 15 kg load on H-grade. It was also found that the EER values increased almost linearly with the increase in load and grade.

(b) Carrying loads on head

- On H-grade (horizontal grade) the mean EER (range) was

Table 4.16: Physiological response in terms of energy expenditure rate during various modes of carrying loads

Modes of carrying loads	*Energy Expenditure Rate (keal/min) Load (kg)*					
	15		*20*		*25*	
	Mean and Range	*Increase in EER and Range*	*Mean and Range*	*Increase in EER and Range*	*Mean and Range*	*Increase in EER and Range*
Back						
H	3.11 2.97-3.25	2.00 1.87-2.15	3.58 3.27-3.72	2.48 2.17-2.62	3.90 3.50-4.06	2.80 2.4-2.96
G_1	3.65 3.32-4.33	2.56 2.22-3.23	4.04 3.83-4.43	2.94 2.73-3.33	4.39 3.75-4.74	3.29 2.65-3.64
G_2	4.53 3.91-4.86	3.43 2.81-3.76	4.76 4.26-5.17	3.66 3.16-4.07	5.44 5.25-5.83	4.34 4.15-4.73
Head						
H	3.39 3.18-3.51	2.29 2.08-2.41	4.07 3.45-4.45	2.97 2.35-3.35	4.26 3.77-4.49	4.36 2.67-3.39
G_1	4.28 3.58-4.48	3.18 2.48-3.38	4.72 4.52-4.89	3.62 3.42-3.79	5.17 4.82-5.67	4.07 3.72-4.57
G_2	5.01 4.08-5.49	3.91 2.98-4.39	5.60 4.59-6.42	4.50 3.49-5.32	6.37 5.77-6.92	5.27 4.67-5.82
Shoulder						
H	4.15 3.51-4.62	3.05 2.41-3.52	4.51 3.73-4.95	3.41 2.63-3.85	5.07 4.84-5.17	3.97 3.74-4.07
G_1	4.65 3.74-5.03	3.55 2.64-3.93	5.31 5.08-5.53	4.21 3.98-4.43	5.86 5.56-6.28	4.76 4.46-5.18
G_2	5.54 4.56-6.00	4.44 3.46-4.90	6.50 5.67-7.18	5.4 4.57-6.08	7.17 6.41-7.57	6.07 5.31-6.47

Resting mean EER =1.107 kcal/min, No. of subjects=7
H = Zero grade (Horizontal surface), G_1 = 5% Grade, G_2 = 10% grade

3.39 (3.18-3.51), 4.07 (3.45-4.45) and 4.26 (3.77-4.49) kcal/min and increase in EER (range) from resting level was 2.29 (2.08-2.41), 2.97 (2.35-3.35) and 4.36 (2.67-3.39) kcal/min with 15, 20 and 25 kg loads, respectively. The EER ranged from 3.18-4.49 kcal/min and increase from its resting level ranged from 2.08-3.39 kcal/min.

- On G_1-grade the mean (range) EER was 4.28 (3.58-4.48), 4.72 (4.52-4.89) and 5.17 (4.82-5.67) kcal/min and increase in EER (range) from its resting level was 3.18 (2.48-3.38), 3.62 (3.423.79) and 4.07 (3.72-4.57) kcal/min with 15, 20 and 25 kg loads, respectively. The EER ranged from 3.58-5.67 kcal/min and the increase from its resting level ranged from 2.48-4.57 kcal/min.
- On G_2-grade the mean EER (range) was 5.01 (4.08-5.49), 5.60 (4.59-6.42) and 6.37 (5.77-6.92) kcal/min and increase in EER (range) from its resting level was 3.91 (2.98-4.39), 4.50 (3.49-5.32) and 5.27 (4.67-5.82) kcal/min with 15, 20 and 25 kg loads, respectively. The EER ranged from 4.08-6.92 kcal/min and increase from its resting level ranged from 2.98-5.82 kcal/min.

While considering EER values of carrying different loads on head at different grades, it is evident that the highest value of 6.37 (5.77-6.92) kcal/min was found while carrying 25 kg load on G_2 grade where as lowest value of 3.39 (3.18-3.51) kcal/min was found for carrying 15 kg load on H-grade. It was also found that the EER values increased almost linearly with increase in load and grade.

(c) Carrying loads on shoulder

- On H-grade the mean EER (range) was 4.15 (3.51-4.62), 4.51 (3.73-4.95) and 5.07 (4.84-5.17) kcal/min and increase in EER (range) from its resting level was 3.05 (2.41-3.52), 3.41 (2.63-3.85) and 3.97 (3.74-4.07) kcal/min with 15, 20 and 25 kg loads, respectively. The EER ranged from 3.51-5.17 kcal/min and increase from resting level ranged from 2.41-4.07 kcal/min.
- On G_1-grade the mean (range) EER was 4.65 (3.74-5.03), 5.31 (5.08-5.53) and 5.86 (5.56-6.28) kcal/min and increase in EER (range) from its resting level was 3.55 (2.64-3.93), 4.21 (3.98-4.43) and 4.76 (4.46-5.18) kcal/min with 15, 20 and 25 kg loads, respectively. The EER ranged from 3.74-6.28 kcal/min and increase from resting level ranged from 2.64-5.18 kcal/min.

- On G_2-grade the mean EER (range) was 5.54 (4.56-6.00), 6.50 (5.67-7.18) and 7.17 (6.41-7.57) and increase in EER (range) from its resting level was 4.44 (3.46-4.90), 5.40 (4.57-6.08) and 6.07 (5.31-6.47) kcal/minute with 15, 20 and 25 kg loads, respectively. The EER ranged from 4.56-7.57 and increase from resting level ranged from 3.46-6.47 kcal/min.

While considering EER values of carrying different loads on shoulder at different grades, it is evident that the highest value of 7.17 (6.41-7.57) kcal/min was found for carrying 25 kg load on G_2 grade whereas lowest value of 4.15 (3.51-4.62) kcal/min for carrying 15 kg load on H-grade. It was also found that the EER values increased almost linearly with the increase in load and grade. Interaction effect of load, mode and grade was found to be non significant

As per the classification of the activities given by Christensen (1953), carrying loads on back at different grades where EER ranged between 3.11- 5.44 kcal/min was classified as light to moderate activities whereas carrying loads on head at different grades where the EER ranged in between 3.39-6.37 kcal/min was classified as light to heavy activities. However, carrying loads on shoulder at different grades where the EER ranged in between 4.15- 7.17 kcal/min was classified as moderate to heavy activities.

To conclude, it can be admitted that the highest mean value of EER i.e. 7.17 (6.41-7.57) kcal/min was found for carrying 25 kg load on shoulder at G_2-grade which was more than two times demanding physiologically than the lowest mean value of 3.11 (2.97-3.25) kcal/min obtained while carrying 15 kg load on the back at H-grade. The carrying loads on shoulder was more demanding in terms of EER than the head and back modes for all the grades. Moreover, 25 kg load and G_2-grade produced greatest metabolic load than other two loads and grades. Hence, the back mode of carrying loads was physiologically (EER) least demanding than the other two modes of head and shoulders.

Oxygen consumption rate per kg body weight

Absolute VO_2 is simply the maximal oxygen uptake level which one could reach during the test and is expressed in liters per minute (l/min). It is obvious that a 80 kg person will have a larger O_2 uptake than a 40 kg person. Therefore, VO_2 is divided by body weight (ml/min/kg) to express VO_2 relative to body weight.

Table 4.17 shows the OCR/kg body weight values for different modes of carrying loads at different grades. Analysis of variance is shown in Appendix B, Table B-5. The mean OCR/kg at resting level is 3.781 ml/min/kg.

(a) Carrying loads on back

- On H-grade the mean (range) OCR/kg body weight was 10.84 (10.16-12.63), 12.59 (11.43-13.24) and 13.34 (11.96-13.88) ml/min/kg with 15, 20 and 25 kg loads, respectively.
- On G_1-grade the mean (range) OCR/kg body weight was 12.51 (11.54-14.79), 14.23 (13.34-15.58) and 15.36 (14.52-16.21) ml/min/kg with 15, 20 and 25 kg loads, respectively.
- On G_2-grade the mean (range) OCR/kg body weight was 15.77 (13.61-16.60), 16.77 (14.82-19.01) and 18.61 (17.93-19.94) ml/min/kg with 15, 20 and 25 kg loads, respectively.

(b) Carrying loads on head

- On H-grade the mean (range) OCR/kg body weight was 11.84 (11.30-13.64), 14.55 (12.48-16.90) and 15.23 (14.29-17.47) ml/min/kg with 15, 20 and 25 kg loads, respectively.
- On G_1-grade the mean (range) OCR/kg body weight was 14.93 (12.46-17.12), 16.86 (15.45-19.00) and 18.53 (16.80-22.06) ml/min/kg with 15, 20 and 25 kg loads, respectively.
- On G_2-grade the mean (range) OCR/kg body weight was 17.48 (14.21-19.34), 20.06 (15.96-22.88) and 22.76 (20.40-24.82) ml/min/kg with 15, 20 and 25 kg loads, respectively.

(c) Carrying loads on shoulder

- On H-grade the mean (range) OCR/kg body weight was

Table 4.17: Physiological response in terms of oxygen consumption rate per kg body weight during various modes of carrying loads

Modes of carrying loads	*Mean and range OCR per kg body weight (ml/min/kg) Load (kg)*		
	15	*20*	*25*
Back			
H	10.84 10.16-12.63	12.59 11.43-13.24	13.34 11.96-13.88
G_1	12.51 12.51-14.79	14.23 13.34-15.58	15.36 14.52-16.21
G_2	15.77 13.61-16.60	16.77 14.82-19.01	18.61 17.93-19.94
Head			
H	11.84 11.30-13.64	14.55 12.48-16.90	15.23 14.29-17.47
G_1	14.93 12.46-17.12	16.86 15.45-19.00	18.53 16.80-22.06
G_2	17.48 14.21-19.34	20.06 15.96-22.88	22.76 20.40-24.82
Shoulder			
H	14.50 12.42-16.32	16.11 15.11-19.24	17.32 16.53-17.84
G_1	16.24 13.02-18.47	18.97 17.38-21.49	20.03 19.00-21.47
G_2	19.34 15.86-21.12	23.11 20.06-24.81	24.50 21.91-25.86

Resting mean OCR/kg body weight =3.781 ml/min/kg, No. of subjects=7
H = Zero grade (Horizontal surface), G_1 = 5% Grade, G_2 = 10% grade

14.50 (12.42-16.32), 16.11 (15.11-19.24) and 17.32 (16.53-17.84) ml/min/kg with 15, 20 and 25 kg loads, respectively.

- On G_1-grade the mean (range) OCR/kg was 16.24 (13.02-18.47), 18.97 (17.38-21.49) and 20.03 (19.00-21.47) ml/min/kg with 15, 20 and 25 kg loads, respectively.
- On G_2-grade the mean (range) OCR/kg body weight was 19.34 (15.86-21.21), 23.11 (20.06-24.81) and 24.50 (21.91-25.86) ml/min/kg with 15, 20 and 25 kg loads, respectively.
- The interaction effect of load, mode and grade was found to be non-significant.

The data revealed that the highest OCR/kg body weight of 24.50 (21.91-25.86) ml/min/kg was found while carrying 25 kg load on shoulder at G_2 grade whereas lowest value of 10.84 (10.16-12.63) ml/min/kg was found for carrying 15 kg load on back at H-grade. Considering the OCR/kg body weight value, carrying loads on shoulder was physiologically more demanding than the other modes.

METS Metabolic equivalents VO_2/kg/min

Table 4.18 revealed the METS values for different modes of carrying loads at different grades. Analysis of variance is shown in Appendix B, Table B-6. The mean value of METS at resting level is 1.08 Vo_2/kg/min.

(a) Carrying loads on back

- On H-grade the mean METS was 3.09 (2.90-3.60), 3.59 (3.26-3.78) and 3.81 (3.41-3.88) Vo_2/kg/min with 15, 20 and 25 kg loads, respectively.
- On G_1-grade the mean METS was 3.58 (3.24-4.22), 4.06 (3.81-4.45) and 4.38 (4.15-4.63) Vo_2/kg/min with 15, 20 and 25 kg loads, respectively.
- On G_2-grade the mean METS was 4.5 (3.89-4.83), 4.79 (4.23-5.43) and 5.31 (5.12-5.69) Vo_2/kg/min with 15, 20 and 25 kg loads, respectively.

(b) Carrying loads on head

- On H-grade the mean METS was 3.39 (3.16-3.89), 4.15 (3.56-4.82) and 4.35 (4.08-4.99) Vo_2/kg/min with 15, 20 and 25 kg loads, respectively.
- On G_1-grade the mean METS was 4.26 (3.56-4.89), 4.81 (4.41-5.43) and 5.29 (4.94-6.30) Vo_2/kg/min with 15, 20 and 25 kg loads, respectively.
- On G_2-grade the mean METS was 4.99 (4.06-5.52), 5.73 (4.56-6.53) and 6.50 (5.83-6.76) Vo_2/kg/min with 15, 20 and 25 kg loads, respectively.

Table 4.18: Metabolic equivalents during various modes of carrying loads

Modes of carrying loads	*Metabolic equivalents (VO_2/kg/min) Load (kg)*		
	15	*20*	*25*
Back			
H	3.09 2.90-3.60	3.59 3.26-3.78	3.81 3.41-3.88
G_1	3.58 3.24-4.22	4.06 3.81-4.45	4.38 4.15-4.63
G_2	4.5 3.89-4.83	4.79 4.23-5.43	5.31 5.12-5.69
Head			
H	3.39 3.16-3.89	4.15 3.56-4.82	4.35 4.08-4.99
G_1	4.26 3.56-4.89	4.81 4.41-5.43	5.29 4.94-6.30
G_2	4.99 4.06-5.52	5.73 4.56-6.53	6.50 5.83-6.76
Shoulder			
H	4.14 3.48-4.60	4.60 4.34-5.49	4.95 4.72-5.09
G_1	4.64 3.72-5.27	5.42 4.96-6.14	5.72 5.43-6.13
G_2	5.52 4.53-6.03	6.60 5.73-7.09	7.00 6.26-7.39

Resting METS= 1.08 VO_2/kg/min, No. of subjects=7
H= Zero grade (Horizontal surface), **G_1**=5 % Grade, **G_2**=10% grade **H**=Zero grade (Horizontal surface), **G_1**=5% Grade, **G_2**=10% grade

(c) Carrying loads on shoulder

- On H-grade the mean METS was 4.14 (3.48-4.60), 4.60 (4.34-5.49) and 4.95 (4.72-5.09) Vo_2/kg/min with 15, 20 and 25 kg loads, respectively.
- On G_1-grade the mean METS was 4.64 (3.72-5.27), 5.42 (4.96-6.14) and 5.72 (5.43-6.13) Vo_2/kg/min with 15, 20 and 25 kg loads, respectively.

- On G_2-grade the mean METS were 5.52 (4.53-6.03), 6.60 (5.73-7.09) and 7.00 (6.26-7.39) Vo_2/kg/min with 15, 20 and 25 kg loads, respectively. The interaction effect of load, mode and grade (F= 0.787 ns) was found to be non-significant.

The data revealed that the highest METS was observed during carrying loads on shoulder with all the loads and grades. The maximum METS was found during carrying 25 kg load on shoulder on G_2-grade and the lowest was found during carrying 15 kg load on back at H-grade.

Considering the effect of interaction of three main factors, it can be concluded that though the interaction effect was found non-significant for all the physiological parameters, the values were much higher than the normal. The lowest values for five physiological parameters were found when carrying load on back. The highest values for all the loads, modes and grades were observed while carrying 25 kg load on shoulder at G_2-grade whereas the lowest values were found during carrying 15 kg load on head at H-grade.

Rating of perceived exertion

To study the effect of the load, mode and grade on the subjects' perception of exertion, they were asked to give ratings on a scale of 0-10 after each trial. Table 4.19 shows rating of perceived exertion and is depicted in Fig. 4.8. The analysis of variance is shown in Appendix B, Table B-7.

It was found that the mean rating of perceived exertion ranged from 1.85-8.85. When comparing the ratings with verbal expression it was found that the subjects felt weak to almost maximum exertion while carrying various loads on different grades.

It was found that while carrying loads on the back, the ratings ranged from 1.85 to 5.71 which corresponded to as weak to strong exertion. On H-grade the RPE score was 1.85, 2.28 and 3.28, for G_1-grade it was 2.28, 3.00 and 4.71 and for G_2-grade it was 2.71, 4.28 and 5.71 for 15, 20 and 25 kg loads, respectively.

It was found that while carrying loads on head, RPE score ranged from 3.71-7.14. When comparing the rating with verbal expression, the activities fell under some what strong to strong exertion category. On H-grade the RPE score was 3.71, 4.02 and 5.57, for G_1-grade it was 4.28, 4.42 and 7.00 and on G_2- grade it was 4.57, 6.14 and 7.14 with 15, 20 and 25 kg loads, respectively.

It was found that while carrying loads on shoulder, the RPE score ranged from 5.28-8.85 which revealed heavy to almost maximum exertion. On H-grade the RPE score was 5.28, 5.54 and 7.71, for G_1-grade it was 6.14, 6.67 and 7.85 for G_2-grade it was 7.00, 7.42 and 8.85 with 15, 20 and 25 kg loads, respectively.

Table 4.19: Rating of perceived exertion during various modes of carrying loads

Modes of carrying loads	*RPE score Load (kg)*		
	15	*20*	*25*
Back			
H	1.85	2.28	3.28
G_1	2.28	3.00	4.71
G_2	2.71	4.28	5.71
Head			
H	3.71	4.02	5.57
G_1	4.28	4.42	7.00
G_2	4.57	6.14	7.14
Shoulder			
H	5.28	5.54	7.71
G_1	6.14	6.67	7.85
G_2	7.00	7.42	8.85

Rating Scale

0.0 Nothing at all
0.5 Extremely weak
1.0 Very weak
2.0 Weak
3.0 Moderate
4.0 Some what strong
5.0 Heavy

6.0 Strong
7.0 Very Strong
8.0 Extremely Strong
9.0 Almost Maximum
10.0 Maximum

H=Zero grade (Horizontal surface), $\mathbf{G_1}$=5% Grade, $\mathbf{G_2}$=10% grade

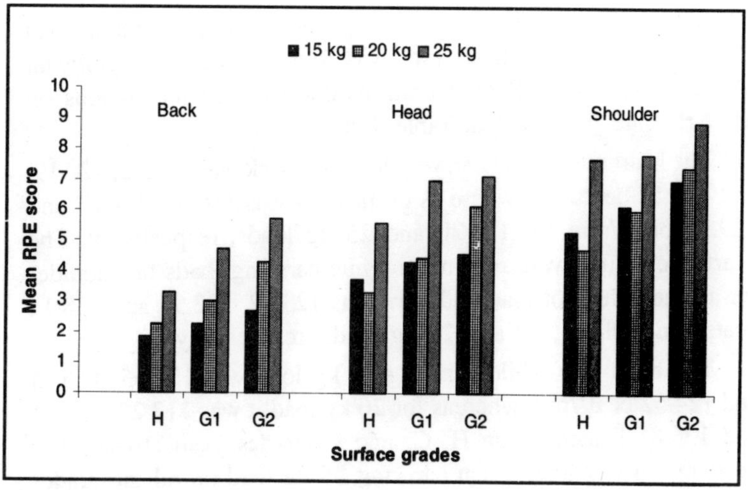

Fig. 4.8 : Variation in rating of perceived exertion for different modes of carrying loads on various surface grades

It was found that carrying any amount of load on shoulder at any of the grades was perceived as in-between the heavy to almost maximum exertion category. The interaction effect of load, mode and grade ($p<0.01$) was found to be significant

It was concluded from the above results that the highest RPE score of 8.85 was found during carrying 25 kg load at G_2-grade whereas the lowest score of 1.85 was found during carrying 15 kg load at H-grade. It was also found that the RPE scores increased with increase in load and grade. However, the back mode of carrying loads was found to produce lowest RPE score in comparison to other modes at all grades.

11. Effect of Interaction of Main Factors

Three interactions of main factors were (load and mode), (load and grade) and (mode and grade) and their effects on physiological responses and perceived exertion were studied. For interaction effect of two factors, one factor was eliminated.

Effect on heart rate

Interaction effect of (load and mode), (mode and grade) and (load and grade) for mean heart rate was found to be non-significant (Appendix B, Table B-1). Variations due to three interactions on heart rate are presented in Table 4.20 .

The heart rate during carrying loads on back was 112.83, 120.19, and 124.52 beats/min where as on head, it was 116.30, 124.67 and 129.30 beats/min for 15, 20 and 25 kg loads, respectively. The heart rate values were maximum while carrying loads on shoulder for all the values of load and varied as 123.42, 132.95 and 138.14 beats/min with 15, 20 and 25 kg loads, respectively.

The heart rate while carrying 15 kg load was 109.65, 116.65 and 124.52 beats/min whereas for 20 kg load it was 117.25, 123.71 and 136.85 beats/min on H, G_1 and G_2-grades, respectively. The heart rate was higher when carrying 25 kg load on all the grades and varied as 121.20, 129.42 and 141.34 beats/min on H, G_1 and G_2-grades, respectively.

Similarly, the heart rate while carrying loads on H-grade was 111.26, 115.03 and 121.80 beats/min whereas on G_1-grade, it was 117.25, 122.19 and 129.85 beats/min on back, head and shoulder modes, respectively. However, for G_2-grade the heart rate values were higher than other two grades and were in the order of 121.20, 129.42 and 141.34 beats/min on back, head and shoulder modes, respectively.

It could be seen that the lowest heart rate of 109.65 beats/min was found when carrying 15 kg load on the back at H-grade whereas highest value of 142.85 beats/min was found while carrying loads on shoulder at G_2-grade.

Effect on increase in heart rate from resting level

The interaction effect of (load and mode), (mode and grade) and

Table 4.20 : Heart rate for interaction effect of main factors (load and mode), (load and grade) and (mode and grade)

S. No.	*Load (kg)*	*HR (beats/min)* *Mode*		
		Back	*Head*	*Shoulder*
1	15	112.83	116.30	123.42
2	20	120.19	124.67	132.95
3	25	124.52	129.30	138.14
		Load (kg)		
	Grade	*15*	*20*	*25*
1	H (Zero grade)	109.65	117.25	121.20
2	G_1 (5 % Grade)	116.65	123.71	129.42
3	G_2 (10 % grade)	126.25	136.85	141.34
		Grade (%)		
	Mode	*H*	G_1	G_2
1	Back	111.26	117.75	128.53
2	Head	115.03	122.19	133.07
3	Shoulder	121.80	129.85	142.85
	SEm±		1.545	
	C.D. at 5 %		4.316	

(load and grade) for increase in heart rate from resting level was also found to be non-significant (Appendix B, Table B-2). Variations due to three interactions on increase in heart rate are presented in Table 4.21.

The increase in heart rate from its resting level while carrying loads on the back was 35.95, 42.85 and 47.52 beats/min while on the head it was 39.04, 47.42 and 52.04 beats/min with 15, 20 and 25 kg loads, respectively. However, when carrying loads on shoulder the maximum increase in heart rate was observed as 46.19, 55.80 and 60.90 beats/min with 15, 20 and 25 kg loads, respectively.

The increase in heart rate while carrying 15 kg load was 32.42, 39.47 and 49.28 beats/min whereas with 20 kg load it was 40.00, 46.47 and 59.61 beats/min on H, G_1 and G_2 grades, respectively.

However, thc highest increase was observed while carrying 25 kg load and it was 43.90, 52.28 and 64.28 beats/min on H, G_1 and G_2-grades, respectively.

The increase in heart rate while carrying loads on H-grade was 34.00, 37.76 and 44.57 whereas on G_1-grade, it was 40.52, 45.00 and 52.71 beats/min on back, head and shoulder, respectively. However, the highest increase was found for G_2-grade as 51.80, 55.76 and 65.61 beats/min on back, head and shoulder, respectively

Table 4.21: Increase in heart rate for interaction effect of main factors (load and mode), (load and grade) and (mode and grade)

S. No.	*Load (kg)*	*Increase in HR (beats/min)* *Mode*		
		Back	*Head*	*Shoulder*
1	15	35.95	39.04	46.19
2	20	42.85	47.42	55.80
3	25	47.52	52.04	60.90
			Load (kg)	
	Grade	*15*	*20*	*25*
1	H (Zero grade)	32.42	40.00	43.90
2	G_1 (5 % Grade)	39.47	46.47	52.28
3	G_2 (10 % grade)	49.28	59.61	64.28
			Grade (%)	
	Mode	*H*	*G_1*	*G_2*
1	Back	34.00	40.52	51.80
2	Head	37.76	45.00	55.76
3	Shoulder	44.57	52.71	65.61
	SEm±		1.463	
	C.D. at 5 %		4.087	

Effect on oxygen consumption rate

Interaction effect of load and mode was found to be non-significant but the other two interaction effects i.e. mode and grade ($p<0.05$) and load and grade ($p<0.01$) were found to be significant (Appendix B, Table B-3). Variations due to three interactions on OCR are presented in Table 4.22 .

The OCR values while carrying load on the back was 0.746, 0.818 and 0.908 l/min where as it was 0.837, 0.950 and 1.044 l/min with 15, 20 and 25 kg loads, respectively. The highest values of OCR were found while carrying load on shoulder as 0.948, 1.079 and 1.196 l/min with 15, 20 and 25 kg loads, respectively.

Table 4.22: OCR for interaction effect of main factors (load and mode), (load and grade) and (mode and grade)

S. No.	*Load (kg)*	*OCR (l/min) Mode*		
		Back	*Head*	*Shoulder*
1	15	0.746	0.837	0.948
2	20	0.818	0.950	1.079
3	25	0.908	1.044	1.196
		Load (kg)		
	Grade	*15*	*20*	*25*
1	H (Zero grade)	0.703	0.806	0.874
2	G_1 (5 % Grade)	0.831	0.929	1.019
3	G_2 (10 % grade)	0.997	1.112	1.254
		Grade (%)		
	Mode	*H*	*G_1*	*G_2*
1	Back	0.700	0.799	0.974
2	Head	0.774	0.935	1.121
3	Shoulder	0.910	1.045	1.268
	SEm±		0.014	
	C.D. at 5 %		0.041	

The OCR values while carrying 15 kg load was 0.703, 0.831 and 0.997 l/min whereas with 20 kg load it was 0.806, 0.929 and 1.112 l/min for H, G_1 and G_2 grades, respectively. However, the highest OCR values were observed while carrying 25 kg load as 0.874, 1.019 and 1.254 l/min for H, G_1 and G_2 grades, respectively.

The OCR values while carrying load on back was 0.700, 0.799 and 0.974 l/min whereas on head it was 0.774, 0.935 and 1.121 l/min for H, G_1 and G_2 grades, respectively. However, on shoulder the OCR values were 0.910, 1.045 and 1.268 for H, G_1 and G_2 grades, respectively.

It could, therefore be concluded that the highest value of 1.254 l/min was found while carrying 25 kg load on G_2-grade whereas lowest value was found when carrying 15 kg load on H-grade.

Effect on energy expenditure rate

Interaction effect of load and mode was found to be non-significant but the other two interaction effects i.e. mode and grade ($p<0.05$) and load and grade ($p<0.01$) were found to be significant (Appendix B, Table B-4). Variations due to three interactions on EER are presented in Table 4.23.

The EER values during carrying loads on the back were 3.767, 4.132 and 4.585 kcal/min whereas on the head it was 4.226, 4.797 and 5.270 kcal/min with 15, 20 and 25 kg loads, respectively. The highest values of EER were found while carrying loads on shoulder as 4.785, 5.439 and 6.037 kcal/min with 15, 20 and 25 kg loads, respectively.

The EER values while carrying 15 kg load was 3.552, 4.194 and 5.032 kcal/min whereas for carrying 20 kg load it was 4.057, 4.691 and 5.619 kcal/min on H, G_1 and G_2 grades, respectively. The highest EER values were observed while carrying 25 kg load in the order of 4.413, 5.148 and 6.331 kcal/min on H, G_1 and G_2 grades, respectively.

The EER values while carrying load on H-grade was 3.534, 3.909 and 4.579 kcal/min while on G_1-grade it was 4.032, 4.724 and 5.277 kcal/min on back, head and shoulders, respectively. The highest EER values were found on G_2-grade in the order of 4.917, 5.660 and 6.404 kcal/min on back, head and shoulders, respectively.

Table 4.23: EER for interaction effect of main factors (load and mode), (load and grade) and (mode and grade)

S. No.	Load (kg)	EER (kcal/min) Mode		
		Back	*Head*	*Shoulder*
1	15	3.767	4.226	4.785
2	20	4.132	4.797	5.439
3	25	4.585	5.270	6.037
		Load (kg)		
	Grade	*15*	*20*	*25*
1	H (Zero grade)	3.552	4.057	4.413
2	G_1 (5 % Grade)	4.194	4.691	5.148
3	G_2 (10 % grade)	5.032	5.619	6.331
		Grade (%)		
	Mode	*H*	G_1	G_2
1	Back	3.534	4.032	4.917
2	Head	3.909	4.724	5.660
3	Shoulder	4.579	5.277	6.404
	SEm±		0.075	
	C.D. at 5 %		0.210	

Effect on OCR/kg body weight

For OCR/kg interaction effect of (load and mode) and (mode and grade) were found to be non-significant but the interaction effect of load and grade ($p<0.01$) was found to be significant (Appendix B, Table B-5). Variations due to interaction effects are presented in Table 4.24

The OCR values per kg body weight while carrying loads on the back were 13.045, 14.535 and 15.775 ml/min/kg whereas on the head it was 14.753, 17.160 and 18.848 ml/min/kg with 15, 20 and 25 kg loads, respectively. The highest values were observed while carrying loads on shoulder which were 16.699, 19.406 and 20.624 ml/min/kg with 15, 20 and 25 kg loads, respectively.

The OCR values per kg body weight while carrying 15 kg load were 12.397, 14.564 and 17.536 ml/min/kg whereas with 20 kg load, it was 14.423, 16.692 and 19.986 ml/min/kg on H, G_1 and G_2 grades, respectively. The highest values were observed with 25 kg load which were 15.305, 17.972 and 21.962 ml/min/kg on H, G_1 and G_2 grades, respectively.

The OCR values per kg body weight while carrying load on H-grade was 12.262, 13.879 and 15.984 ml/min/kg whereas on G_1-grade it was 14.040, 16.776 and 18.419 ml/min/kg on back, head and shoulder, respectively. The highest values were found on G_2-grade which was 17.051, 20.107 and 22.326 ml/min/kg on back, head and shoulder, respectively.

Table 4.24: OCR/kg body weight for interaction effect of main factors (load and mode), (load and grade) and (mode and grade)

S. No.	*Load (kg)*	*OCR/kg body weight (ml/min/kg)* Mode		
		Back	*Head*	*Shoulder*
1	15	13.045	14.753	16.699
2	20	14.535	17.160	19.406
3	25	15.775	18.848	20.624
		Load (kg)		
	Grade	*15*	*20*	*25*
1	H (Zero grade)	12.397	14.423	15.305
2	G_1 (5 % Grade)	14.564	16.692	17.972
3	G_2 (10 % grade)	17.536	19.986	21.962
		Grade (%)		
	Mode	*H*	*G_1*	*G_2*
1	Back	12.262	14.040	17.051
2	Head	13.879	16.776	20.107
3	Shoulder	15.984	18.419	22.326
	SEm±		0.296	
	C.D. at 5 %		0.828	

Effect on metabolic equivalents

The interaction effect of load and mode was found to be non-significant but the interaction effects of mode and grade ($p<0.05$) and load and grade ($p<0.01$) at was found to be significant (Appendix B, Table B-6). Variations due to interaction effects are presented in Table 4.25.

The METS during carrying loads on the back was 3.726, 4.152 and 4.507 kg/min whereas on the head; it was 4.214, 4.909 and 4.771 VO_2/kg/min with 15, 20 and 25 kg loads, respectively. The highest values were found with the shoulder mode which were 4.771, 5.544 and 5.891 VO_2/kg/min with 15, 20 and 25 kg loads, respectively.

Table 4.25: METS for interaction effect of main factors (load and mode), (load and grade) and (mode and grade)

S. No.	*Load (kg)*	*METS (VO_2/kg/min) Mode*		
		Back	*Head*	*Shoulder*
1	15	3.726	4.214	4.771
2	20	4.152	4.902	5.544
3	25	4.507	5.384	5.891
		Load (kg)		
	Grade	*15*	*20*	*25*
1	H (Zero grade)	3.542	4.119	4.373
2	G_1 (5 % Grade)	4.160	4.769	5.136
3	G_2 (10 % grade)	5.009	5.710	6.273
		Grade (%)		
	Mode	*H*	*G_1*	*G_2*
1	Back	3.503	4.011	4.871
2	Head	3.964	4.793	5.744
3	Shoulder	4.568	5.261	6.378
	SEm±		0.084	
	C.D. at 5 %		0.236	

The METS during carrying 15 kg load was 3.542, 4.160 and 5.009 VO_2/kg/min whereas with 20 kg load; it was 4.119, 4.769 and 5.710 VO_2/kg/min on H, G_1 and G_2 grades, respectively. The highest values were observed with 25 kg load and it was 4.373, 5.136 and 6.273 VO_2/kg/min on H, G_1 and G_2 grades, respectively.

The METS during carrying loads on H-grade was 3.503, 3.964 and 4.568 kg/min whereas on G_1-grade, it was 4.011, 4.793 and 5.261 VO_2/kg/min on the back, head and shoulder modes, respectively. The highest values were observed on G_2-grade as 4.871, 5.744 and 6.378 VO_2/kg/min on the back, head and shoulder modes, respectively.

Effect on rating of perceived exertion

Interaction effect of (load and mode) and (load and grade) was found to be significant ($p<0.01$) but the interaction effects of mode and grade was found to be non-significant (Appendix B, Table B-7). Variations due to interaction effects on RPE are presented in Table 4.26.

The RPE score during carrying loads on the back was 2.285, 3.190 and 4.571, and on head, it was 4.190, 4.619 and 6.571. The highest scores were observed on shoulder mode as 6.142, 6.147 and 8.142 with 15, 20 and 25 kg loads, respectively.

The RPE score during carrying 15 kg load was 3.619, 4.238 and 4.761 whereas with 20 kg load, it was 3.482, 4.476 and 5.952. The highest scores were observed while carrying 25 kg load which were 5.523, 6.523 and 7.238 on H, G_1 and G_2 grades, respectively

The RPE score during on H-grade was 2.476, 4.190 and 5.904 whereas on G_1-grade, it was 3.333, 5.238 and 6.666. The highest scores were observed on G_2-grade as 4.238, 5.952 and 7.761 on the back, head and shoulder, respectively.

It could be inferred that the effects of all the three interactions were found to be non-significant for the heart rate and increase in the heart rate from resting level. However, interaction effect of load and mode was found to be non-significant with all the physiological parameters except for perceived exertion ($p<0.01$). Also the interaction effect of load and grade was found to be significant ($p<0.01$) for all physiological parameters and RPE also except for the heart rate and increase in the heart rate from its resting level. As far as the interaction effect of mode and grade was concerned, it was found to be significant ($p<0.05$) for OCR, EER and METS only.

Table 4.26: RPE for interaction effect of main factors (load and mode), (load and grade) and (mode and grade)

S. No.	*Load (kg)*	*RPE score Mode*		
		Back	*Head*	*Shoulder*
1	15	2.285	4.190	6.142
2	20	3.190	4.619	6.147
3	25	4.571	6.571	8.142
		Load (kg)		
	Grade	15	20	25
1	H (Zero grade)	3.619	3.428	5.523
2	G_1 (5 % Grade)	4.238	4.476	6.523
3	G_2 (10 % grade)	4.761	5.952	7.238
		Grade (%)		
	Mode	H	G_1	G_2
1	Back	2.476	3.333	4.238
2	Head	4.190	5.238	5.952
3	Shoulder	5.904	6.666	7.761
	SEm±		0.084	
	C.D. at 5 %		0.236	

N.B. : Refer Table 4.19 for Rating Scale

12. Effect of Loads, Modes and Grades on Physiological Responses and Perceived Exertion of Subjects

Table 4.27 shows the effect of main factors on physiological cost and perceived exertion. A total of five physiological parameters i.e. Heart Rate (HR), Oxygen Consumption Rate (OCR), Energy Expenditure Rate (EER), OCR per kg body weight, Metabolic Equivalents (METS) as well as Rating of Perceived Exertion (RPE) were selected to observe the variation due to loads, modes and grades.

F-values of ANOVA were computed (Appendix B, Table B-1) to observe the effect of loads, modes and grades on the heart rate and statistically significant differences ($p<0.01$) were observed in respect of all the main factors. It was found that while carrying different loads of 15, 20 and 25 kg the mean heart rate was 117.52, 125.94 and 138.14 beats/min, respectively whereas during various modes of carrying load viz. back, head and shoulder, it was 119.18, 123.43 and 131.50 beats/min, respectively. Similarly, for carrying loads on different grades i.e. H, G_1 and G_2 it was 116.03, 123.26 and 134.82 beats/min, respectively. This reveals that modes and grades have less effect on heart rate of subjects than the amount of load carried by them.

Analysis of variance for increase in mean heart rate (Appendix B, Table B-2) also revealed significant differences ($p<0.01$) regarding loads, modes and grades. The increase in mean heart rate from resting level for carrying different loads was 40.39, 48.69 and 53.49 beats/min, respectively whereas during various modes of carrying load viz. back, head and shoulder, it was 42.11, 46.17 and 53.30 beats/min, respectively and for carrying on different grades i.e. H, G_1 and G_2 it was 38.77, 46.07 and 57.73 beats/min, respectively.

F-values of ANOVA were computed (Appendix B, Table B-3) to know the effect of load, mode and grade on mean OCR values and significant differences ($p<0.01$) were demonstrated in respect of all the main factors. It was found that the mean OCR (l/min) values during carrying 15, 20 and 25 kg loads were 0.843, 0.949

Table 4.27: Effect of main factors i.e. load, mode and grade

S. No.	Main effects	Physiological parameters						
		HR (beats/ min)	*Increase in HR (beats/ min)*	*OCR (l/min)*	*EER (kcal/ min)*	*OCR/kg (ml/kg/ min)*	*METS Vo_2/ (kg/min)*	*RPE score*
	Load (kg)							
1	15	117.52	40.39	0.843	4.25	14.83	4.23	4.20
2	20	125.94	48.69	0.949	4.78	17.03	4.86	4.61
3	25	138.14	53.49	1.049	5.29	18.41	5.26	6.42
	Mode							
1	Back	119.18	42.11	0.824	4.16	14.45	4.12	3.34
2	Head	123.43	46.17	0.943	4.76	16.92	4.83	5.12
3	Shoulder	131.50	53.30	1.074	5.42	18.91	5.40	6.77
	Grade							
1	H (Zero grade)	116.03	38.77	0.795	4.00	14.04	4.01	4.19
2	G_1 (5% grade)	123.26	46.07	0.926	4.67	16.41	4.68	5.07
3	G_2 (10% grade)	134.82	57.73	1.121	5.66	19.82	5.66	5.98
	SEm±	0.892	0.845	0.008	0.043	0.171	0.048	0.071
	C.D. at 5 %	2.492	2.360	0.024	0.121	0.478	0.136	0.200

H=Zero grade (Horizontal surface), G_1=5% Grade, G_2=10% grade

and 1.049 l/min, respectively whereas during various modes of carrying load viz. back, head and shoulder, it was 0.824, 0.943 and 1.074 l/min, respectively. However, for different grades i.e. H, G_1 and G_2 it was 0.795, 0.926 and 1.121 l/min, respectively. The highest value of OCR was found during carrying loads on G_2 grade whereas lowest was found on H-grade.

F-values of ANOVA were computed (Appendix B, Table B-4) to see the effect of load, mode and grade on mean EER (kcal/min) values and significant differences ($p<0.01$) were demonstrated in respect of load, mode and grade. It was found that the mean EER values during carrying 15, 20 and 25 kg loads were 4.25, 4.78 and 5.29 kcal/min, respectively whereas during various modes of carrying load viz. back, head and shoulder, it was 4.16, 4.76 and 5.42 kcal/min, respectively. However, for different grades i.e. H, G_1 and G_2, it was 4.00, 4.67 and 5.66 kcal/min, respectively.

F-values of ANOVA were computed (Appendix B, Table B-5) to observe the effect of load, mode and grade on mean OCR/kg body weight and significant differences ($p<0.01$) were demonstrated regarding load, mode and grade. It was found that the OCR/kg body weight during carrying 15, 20 and 25 kg loads were 14.83, 17.03 and 18.41 ml/min/kg, respectively whereas during various modes of carrying load viz. back, head and shoulder, it was 14.45, 16.92 and 18.91 ml/min/kg, respectively. However, for different grades i.e. H, G_1 and G_2, it was 14.04, 16.41 and 19.82 ml/min/kg, respectively.

F-values of ANOVA were computed to see the effect of load, mode and grade on mean METS and significant differences ($p<0.01$) were demonstrated in respect of load, mode and grade (Appendix B, Table B-6). It was found that METS values during carrying 15, 20 and 25 kg loads were 4.23, 4.86 and 5.26 Vo_2/kg/min whereas during various modes of carrying load viz. back, head and shoulder, it was 4.12, 4.83 and 5.40 VO_2/kg/min, respectively. Similarly, for different grades i.e. on H, G_1 and G_2, it was 4.01, 4.68 and 5.66 VO_2/kg/min, respectively.

F-values of ANOVA were computed (Appendix B, Table B-7) to find out the effect of load, mode and grade on RPE scores and here also significant differences ($p<0.01$) were demonstrated regarding all the main factors. It was found that the RPE score during carrying 15, 20 and 25 kg loads were 4.20, 4.61 and 6.42, respectively whereas during various modes of carrying load viz. back, head and shoulder, it was 3.34, 5.12 and 6.77, respectively. However, for carrying on different grades i.e. H, G_1 and G_2, it was 4.19, 5.07 and 5.98, respectively.

It could, therefore, be inferred that the significant differences were observed regarding main factors i.e. loads, modes and grades for all the five physiological parameters and also, for perceived exertion. The lowest values of physiological responses and perceived exertion were observed while carrying 15 kg load among loads, on zero (H) grade among grades and carrying on back among different modes. The lowest values for carrying on the back may be because of the load mass close to centre of gravity of the body in agreement with Knapik *et al.* (1996).

13. Correlation and Regression Analysis between Various Parameters

Table 4.28 shows the HR, OCR and RPE values of 27 different treatments between which the correlation coefficients were calculated and Table 4.29 shows various regression equations between different parameters and the relationships are shown in Figs. 4.9a and b, Figs. 4.10a and b and Figs. 4.11a and b. The detailed regression analysis is given in Appendix C, Table C-1 to C-6. The heart rate data was significantly ($p<0.01$) correlated with the OCR (r=0.981**) and RPE (r=0.840**) data. Also significant correlation was found between OCR and RPE (r=0.873**).

HR has a significant positive effect on OCR and result indicated (R^2= 0.962) that HR explains 96.20 per cent variation in OCR and vice-versa.

HR has a significant positive effect on RPE and result indicated (R^2= 0.706) that HR explains 70.60 per cent variation in RPE and vice-versa.

OCR has a significant negative effect on RPE whereas RPE has a positive effect on OCR. The result indicated (R^2= 0.761) that OCR explains 76.10 per cent variation in RPE and vice-versa.

14. Regression Analysis between Heart Rate and Independent Variables (grade, mode and load)

Table 4.30 shows the regression equation between heart rate and independent variables (grade, mode and load) and the detailed regression analysis is given in Appendix D, Table D-1. The regression equation between HR and independent variables was $Y = 67.81+9.52X_1+6.09X_2+1.28X_3$ and the coefficient of determination (R^2) was 0.966. This shows that 96.6 per cent variation in the

Table 4.28: Correlation coefficient between various dependent variables

Level of Independent variables				Dependent parameters					
S. No.	Load	Mode	Grade	HR	OCR	RPE	HR	RPE	OCR
1	15	Back	H	105.54	0.62	1.85	105.54	1.85	0.62
2	15	Back	G_1	112.46	0.72	2.28	112.46	2.28	0.72
3	15	Back	G_2	121.64	0.89	2.71	121.64	2.71	0.89
4	15	Head	H	108.67	0.67	3.71	108.67	3.71	0.67
5	15	Head	G_1	116.07	0.84	4.28	116.07	4.28	0.84
6	15	Head	G_2	125.46	0.99	4.57	125.46	4.57	0.99
7	15	Shoulder	H	114.75	0.82	5.28	114.75	5.28	0.82
8	15	Shoulder	G_1	121.42	0.92	6.14	121.42	6.14	0.92
9	15	Shoulder	G_2	134.10	1.09	7.00	134.10	7.00	1.09
10	20	Back	H	112.03	0.71	2.28	112.03	2.28	0.71
11	20	Back	G_1	118.21	0.80	3.00	118.21	3.00	0.80
12	20	Back	G_2	130.32	0.94	4.28	130.32	4.28	0.94
13	20	Head	H	115.85	0.80	3.28	115.85	3.28	0.80
14	20	Head	G_1	122.82	0.93	4.42	122.82	4.42	0.93
15	20	Head	G_2	135.35	1.10	6.14	135.35	6.14	1.10
16	20	Shoulder	H	123.85	0.89	4.71	123.85	4.71	0.89
17	20	Shoulder	G_1	130.10	1.05	6.00	130.10	6.00	1.05
18	20	Shoulder	G_2	144.89	1.29	7.42	144.89	7.42	1.29
19	25	Back	H	116.21	0.77	3.28	116.21	3.28	0.77
20	25	Back	G_1	122.57	0.97	4.71	122.57	4.71	0.97

Table 4.28: *(Contd.)*

21	25	Back	G_2	134.78	1.07	5.71	134.78	5.71	1.07
22	25	Head	H	120.57	0.84	5.57	120.57	5.57	0.84
23	25	Head	G_1	127.67	1.03	7.00	127.67	7.00	1.03
24	25	Head	G_2	139.67	1.26	7.14	139.67	7.14	1.26
25	25	Shoulder	H	126.82	1.00	7.71	126.82	7.71	1.00
26	25	Shoulder	G_1	138.03	1.16	7.85	138.03	7.85	1.16
27	25	Shoulder	G_2	149.57	1.42	8.85	149.57	8.85	1.42
	Correlation coefficient (*r*)			0.981**		0.840**		0.873**	

Load=kg
H=Zero grade (Horizontal surface), $\mathbf{G_1}$=5% Grade, $\mathbf{G_2}$=10% grade

Table 4.29 : Regression analysis between physiological parameters and RPE

S. No.	*Y*	*X*	*a*	*b*	*Regression equation*	*R^2*	*t-value*
1.	OCR	HR	-1.2	0.017	Y= -1.2+0.017 X	0.962	25.40**
2.	HR	OCR	72.02	55.67	Y= 72.02+55.67 X	0.962	25.40**
3.	RPE	HR	-13.25	0.146	Y= -13.25+0.146 X	0.706	7.75**
4.	HR	RPE	100.36	4.80	Y= 100.36+4.80 X	0.706	7.75**
5.	RPE	OCR	-3.12	8.65	Y= -3.12+8.65 X	0.761	8.93**
6.	OCR	RPE	0.879	0.500	Y = 0.879+0.500 X	0.761	8.93**

Note : ** Significant at 1 per cent level of significance.

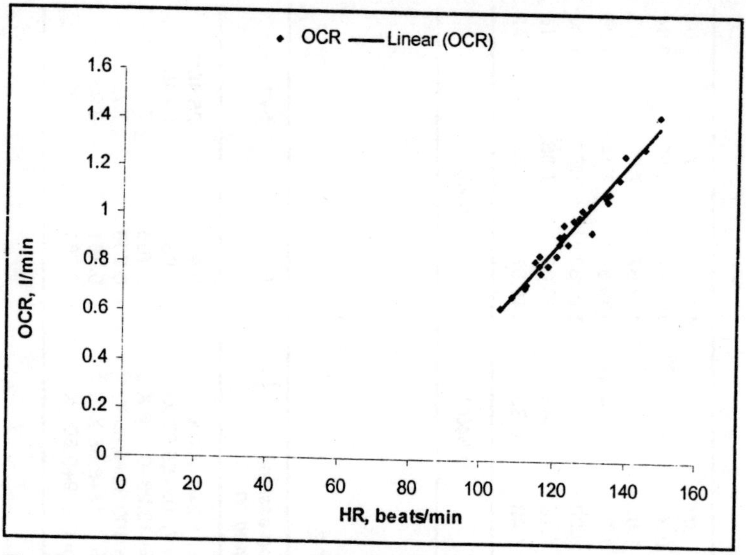

Fig. 4.9a : Relationship between OCR (Y) and HR (X), fitting to the line Y = -1.2+0.017X

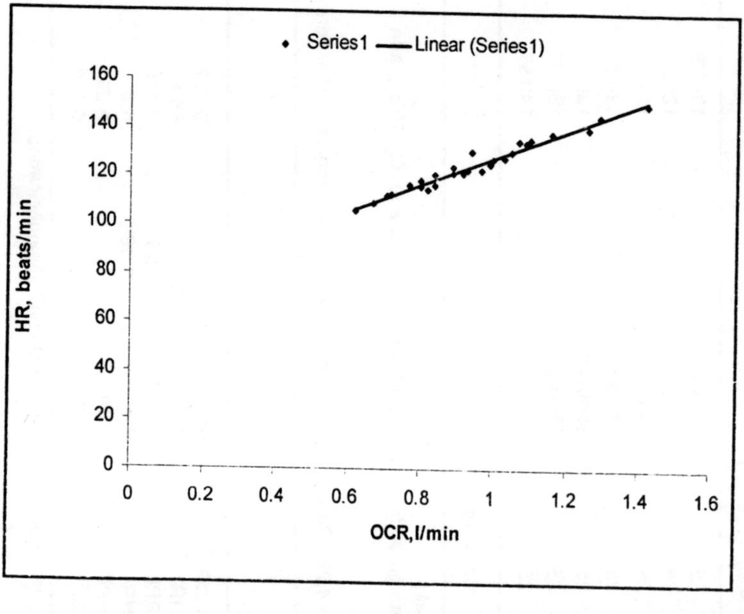

Fig. 4.9b : Relationship between HR (Y) and OCR (X), fitting to the line Y = 72.2+55.67X

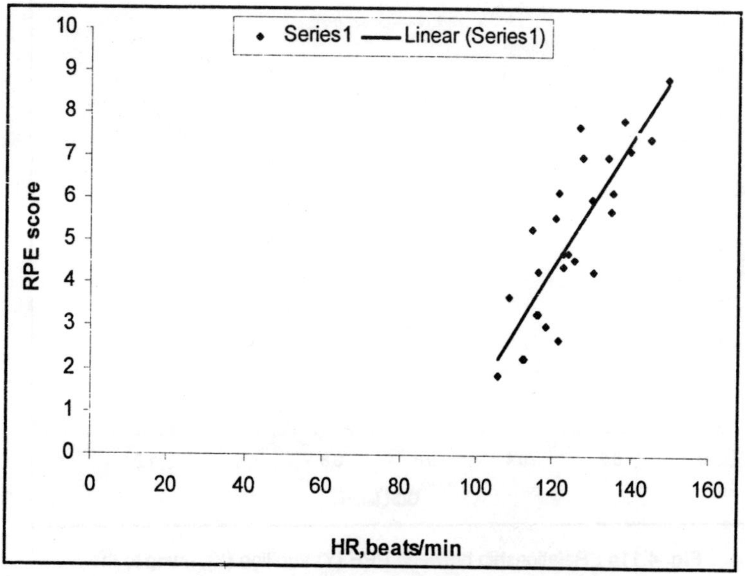

Fig. 4.10a : Relationship between RPE (Y) and HR (X), fitting to the line Y = -13.25+0.146X

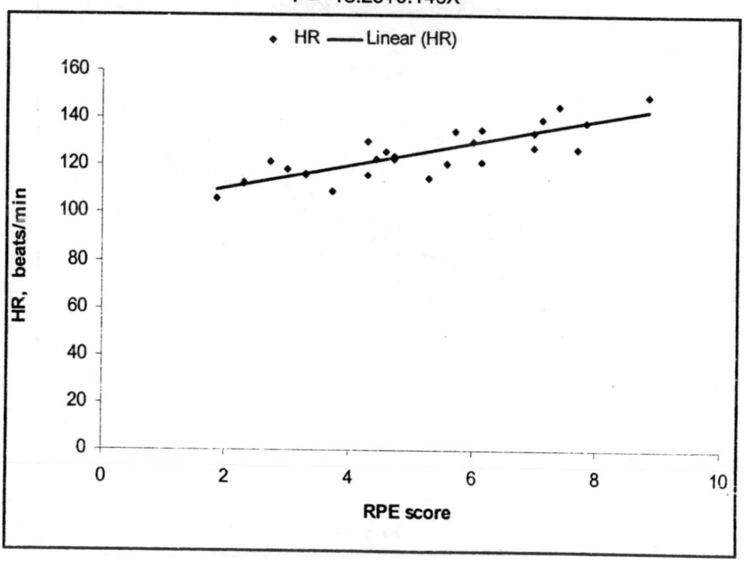

Fig. 4.10b : Relationship between HR (Y) and RPE (X), fitting to the line Y = 100.36+4.80X

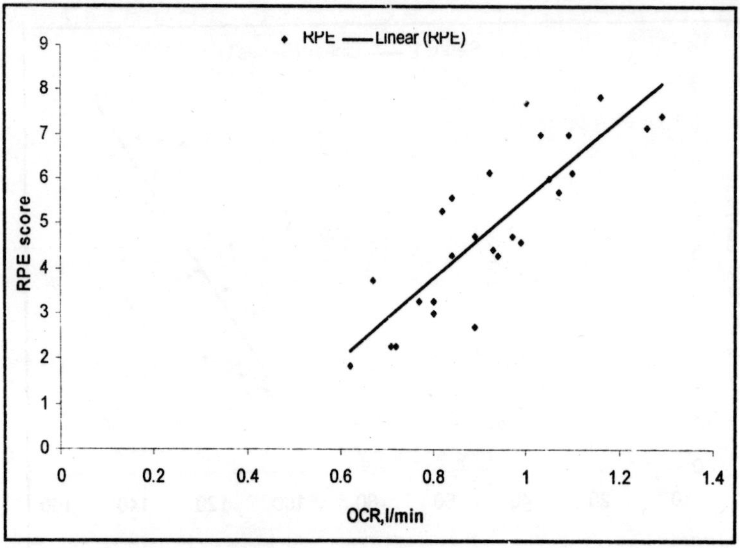

Fig. 4.11a : Relationship between RPE (Y) and line (X), fitting to the
Y = -3.12+8.65X

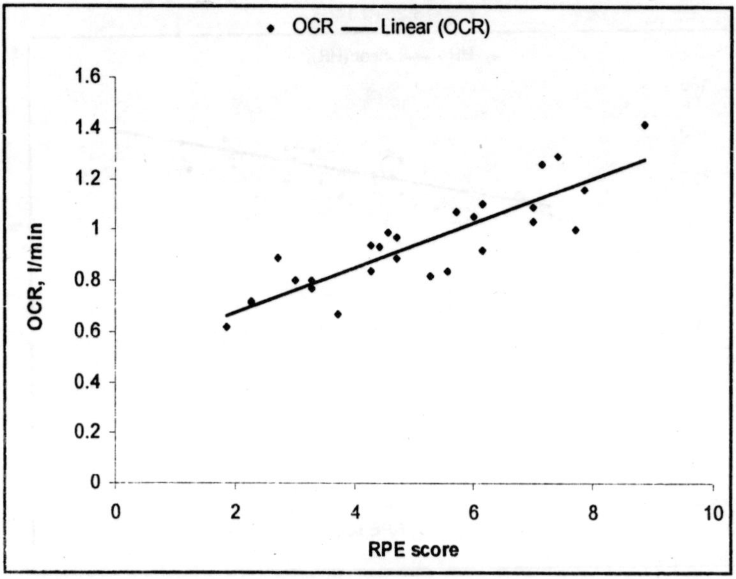

Fig. 4.11b : Relationship between OCR (Y) and RPE (X), fitting to the line
Y = 0.879+0.500X

dependent variable (HR) was due to the combined effect of three independent variables i.e. grade, mode and load. All the three independent variables grade, mode and load have positive significant effect on heart rate.

Table 4.30: Regression analysis between heart rate and independent variables (grade, mode and load)

X	*b*	*t-value*	*a*	*Regression equation*	R^2
1 (Grade)	9.52	18.99**	+67.81	$Y= 67.81+9.52\,X_1+6.09\,X_2+1.28\,X_3$	0.966
2 (Mode)	6.09	12.15**			
3 (Load)	1.28	12.82**			

X_1=1 (H grade), 2 (5% grade), and 3 (10% grade),
X_2=1 (Back), 2 (Head), and 3 (Shoulder)
X_3=15, 20 and 25 kg

Note: Y= Dependent variable, X=Independent variable, a=Intercept, b= Regression coefficient
**= Significant at 1 % level of significance

15. Regression Analysis between Oxygen Consumption Rate and Independent Variables (grade, mode and load)

Table 4.31 shows the regression equation between OCR and independent variables (grade, mode and load) and the detailed regression analysis is given in Appendix D, Table D-2. The regression equation between OCR and the independent variables was $Y = -0.522+0.162X_1+0.119X_2+0.217X_3$ and the coefficient of determination (R^2) was 0.967. This shows that 96.7 per cent variation in the dependent variable (OCR) was due to the combined effect of three independent variables viz. grade, mode and load. All the three independent variables (grade, mode and load) have positive significant effect on oxygen consumption rate.

16. Recommendation on Rest Periods for Women Subjects during Various Modes of Carrying Loads on Different Surface Grades

There are two primary methods which may be used to deal with the

Table 4.31: Regression analysis between oxygen consumption rate and independent variables (grade, mode and load)

X	*b*	*t-value*	*a*	*Regression equation*	R^2
1 (Grade)	0.162	18.60**	-0.522	$Y = -0.522 + 0.162\,X_1 + 0.119X_2 + 0.217\,X_3$	0.967
2 (Mode)	0.119	13.65**			
3 (Load)	0.217	12.44**			

X_1=1(H grade), 2 (5% grade), and 3 (10% grade),
X_2=1 (Back), 2 (Head), and 3 (Shoulder)
X_3=15, 20 and 25 kg

Note: Y= Dependent variable, X= Independent variable, a=Intercept, b= Regression coefficient
**= Significant at 1 % level of significance

negative performance impact of fatigue. The first is adjustment of the job itself in order to delay the on-set of fatigue. The second is through the efficient work-rest policies. It is well known that work-rest schedules influence the output of individuals regardless of whether the task is mental or physical. The first approach requires a focus upon the causes of fatigue while the second method deals with the performance patterns associated with a set of task characteristics. Taylor (1860) experimented with various rest pauses and rest schedules and brought fourth the concept that rest pauses increase efficiency and productivity. Shephard and Walker (1957) have found that if the heavy work is combined with frequent rest conditioned pauses, absenteeism is less than if the work was continuous (Stephen *et al.*, 1984).

Murell (1965) has suggested that regardless of the nature of the task, performance decrements have generally been attributed to the concept of fatigue and introduced a formula for computing the rest period required for a particular kind of activity based on the energy expenditure rate as follow:

$$R = \frac{T\,(K - S)}{(K - 1.5)} \qquad (4.1)$$

Where,

R = Rest required, min.

T = Total working time, min

K = Average kilocalories per minute of work, kcal/min.

S = kcal/min adopted as standard i.e.5 kcal/min.

The constant 1.5 in denominator is an approximation of the resting level in kcal/min.

Table 4.32: Recommendation on rest periods for women subjects during various modes of carrying loads on different surface grades

Modes of carrying loads	*Recommended Rest Period (min.)*					
	On the basis of OCR Load (kg)			*On the basis of HR Load (kg)*		
	15	*20*	*25*	*15*	*20*	*25*
Back						
H	2.07	2.49	2.59	2.75	3.00	3.50
G_1	2.26	3.12	3.36	3.25	3.80	4.00
G_2	2.67	3.38	4.02	3.75	4.00	4.50
Head						
H	2.34	3.02	3.19	3.00	3.5	3.80
G_1	2.54	3.42	3.58	3.50	3.70	4.50
G_2	3.15	4.02	4.17	3.80	4.5	5.25
Shoulder						
H	2.55	3.05	3.15	3.5	4.25	5.25
G_1	3.03	3.3	3.47	4.00	4.75	6.00
G_2	3.5	4.19	4.38	5.00	7.00	8.00

H=Zero grade (Horizontal surface), $\mathbf{G_1}$=5 % Grade, $\mathbf{G_2}$=10 % grade

Table 4.32 shows the time period during which the heart rate and oxygen consumption rate reach to their resting levels. For recommendation on the rest periods, the heart rate and oxygen consumption data was recorded after 8 minutes of activity till the time when responses came down to their resting level. It was found that the OCR values came to their resting level a little before the heart rate values, thus for rest period analysis the heart rate was used as an indicator. However, it took more time by the subjects to mentally prepare themselves for the next combination of activities than the time period required by the responses to come to their resting level.

It was found that while carrying loads on the back at different grades, the heart rate took from 2.00 to 4.50 min to come down to its resting level whereas carrying loads on the head at different grades, it took from 2.34 to 5.25 min. However, while carrying loads on the shoulder at different grades, it took from 2.50-8.00 min after the 8 minutes of activity.

It could, therefore, be concluded that while carrying various loads on the shoulder, the responses took the maximum time to come down to their resting level which was about 8 minutes, whereas carrying loads on the back took the least time in comparison to other two modes.

17. Body Part Discomfort Scores during Various Modes of Carrying Loads by Subjects

Table 4.33 shows the pain and discomfort in various body parts of subjects during carrying loads. However, they were unable to assign score to different combination of trials for different body parts. It was found that subjects felt light to moderate discomfort and pain in various body parts such as lower back and shoulders while carryings loads on the back whereas moderate (15 kg load) to very uncomfortable (20 and 25 kg load) pain in shoulders, hands, neck and palms while carrying loads on shoulder. But, they felt more than moderate discomfort and pain in head and legs while carrying loads on the head. It was also found that the subjects felt moderate discomfort and pain in knees, thighs and toes while carrying loads at G_2-grade.

It can be concluded from the above that carrying loads on the shoulder caused more pain and discomfort in various body parts.

18. Effect of Environment on HR and RPE while Carrying Loads on Back on Horizontal Surface

Humans often exercise strenuously in hot and cold environments

Table 4.33: Body part discomfort and pain during various modes of carrying loads

S. No.	*Load (kg)*	*Mode*	*Grade*	*Pain in body part*	*BPDS*
1.	15	Back	H	Shoulder, lower back	2-3
2.	15	Back	G_1	Shoulder, lower back	
3.	15	Back	G_2	Shoulder, lower back, knees, toes	
4.	15	Head	H	Head, legs	3-4
5.	15	Head	G_1	Head, legs	
6.	15	Head	G_2	Head, legs , knees, toes	
7.	15	Shoulder	H	Shoulders, neck, hands, palms	5-6
8.	15	Shoulder	G_1	Shoulders, neck, hands, palms	
9.	15	Shoulder	G_2	Shoulders, neck, hands, palms, knees, toes	
10.	20	Back	H	Shoulder, lower back	3-4
11.	20	Back	G_1	Shoulder, lower back	
12.	20	Back	G_2	Shoulder, lower back, knees, toes, thighs	
13.	20	Head	H	Head, legs	4-5
14.	20	Head	G_1	Head, legs	
15.	20	Head	G_2	Head, legs, knees, toes	
16.	20	Shoulder	H	Shoulders, neck, hands, palms	6-8
17.	20	Shoulder	G_1	Shoulders, neck, hands, palms	
18.	20	Shoulder	G_2	Shoulders, neck, hands, palms, knees, toes, thighs	
19.	25	Back	H	Shoulder, lower back	3-5
20.	25	Back	G_1	Shoulder , lower back	
21.	25	Back	G_2	Shoulder, lower back, Knees, toes, thighs	
22.	25	Head	H	Head, legs	6-7
23.	25	Head	G_1	Head, legs	
24.	25	Head	G_2	Head, legs, knees, toes	
25.	25	Shoulder	H	Shoulders, neck, hands, palms	8-10
26.	25	Shoulder	G_1	Shoulders, neck, hands, palms	
27.	25	Shoulder	G_2	Shoulders, neck, hands, palms, knees, toes, thighs	

H=Zero grade (Horizontal surface), G_1=5 % Grade, G_2=10 % grade

for reasons of recreation, vocation, and survival. The magnitude of physiological strain imposed by exercise-environmental stress depends on the individual's metabolic rate and capacity for heat exchange with the environment. Muscular exercise increases metabolism by 5 to 15 times the resting rate to provide energy for skeletal muscle contraction. Depending on the type of exercise, 70 to 100 per cent of the metabolism is released as heat and needs to be dissipated in order to maintain body heat balance.

There appears to be general agreement that weather has a profound impact on human health and some of such factors include temperature, humidity, illumination, noise, heat, sunshine, solar radiation, cloud cover, fog etc. The impact of temperature on morbidity and mortality can be assessed at both the seasonal and daily level. The variability in occurrence of numerous illnesses is linked to somewhat predictable seasonal trends in temperature (Persinger, 1980), although significant year-to-year differences do occur. Medical disorders such as bronchitis, peptic ulcer, adrenal ulcer, glaucoma, goiter, eczema, and herpes zoster are related to seasonal variations in temperature (Tromp, 1963). Heart failure and cerebrovascular accidents represent two general mortality categories that have been correlated many times with ambient monthly temperatures (Persinger, 1980). Complications from these disorders can be expected at higher temperatures since the body responds to thermal stress by forcing blood into peripheral areas to promote heat loss through the skin. This increases central blood pressure and encourages constriction of blood vessels near the core of the body. However, increases in heart disease are also noted at very cold temperatures as well. Rapid changes in temperature have been shown to produce a number of physiological changes in the body. Rapid drops may affect blood pH, blood pressure, urination volume, and tissue permeability.

Persinger (1975) found significant negative relationships between relative humidity and "mood scores," which represent a measure of happiness. Rapid changes in the weather often induce a series of negative physiological responses from the body. Humidity has an important impact on mortality since it influences the body's ability to cool itself by means of evaporation of perspiration. In addition, humidity affects human comfort, and the perceived temperature by humans is largely dependent upon atmospheric moisture content (Persinger, 1980) whereas in the summer, high moisture content during hot periods can lessen the body's ability to evaporate perspiration, possibly leading to heat stress. The effects of low humidity can be especially dramatic in winter, when low moisture content induces stress upon the nasal-pharynx and trachea. When very cold, dry air passes through these organs, warming occurs and air temperatures in the pharynx can reach 30ºF. The ability of this warmer air to hold moisture increases dramatically and moisture is extracted at a prodigious rate from the nasal passages and upper respiratory tract, leading to excessive dehydration of these organs (Richards and Marriott, 1974).

Variation due to different environmental conditions in heart rate and RPE is shown in Table 4.34. It was found that the heart rate in cold environment was 104, 109 and 114 beats/min whereas in hot environment it was 108, 113 and 118 beats/min for 15, 20 and 25 kg loads, respectively. It was also found that the activities were perceived to be weak to moderate in cold environment whereas moderate to strong in hot environment. Furthermore, evidence from the scientific literature suggested that the greater portion of the increase in average heart rate in hot and humid environments is more likely to reflect the body's attempts to prevent heat related illness.

Table 4.34: Heart rate and RPE scores in different environments while carrying loads on the back at horizontal surface

S. No.	*Load (kg)*	*Environmental conditions*			
		Cold Temp.:26-29 ºC Humidity: 50-55 %		*Hot Temp.: 35-40 ºC Humidity: 64-67 %*	
		HR (beats/min)	*RPE score*	*HR (beats/min)*	*RPE score*
1.	15	104	1.00	108	3.00
2.	20	109	2.00	113	4.00
3.	25	114	3.00	118	7.00

It can be thus concluded that the different environmental conditions affected the heart rate and perceived exertion. However, in ambient environment the effect on subject's performance might be the result of a combination of different environmental parameters. Thus, there is a need to carry out detailed studies to assess the effect of different environmental conditions (temperature, humidity etc.), in controlled environmental chambers, so that the effect of temperature and humidity could be assessed separately and their limits could be fixed for comfort.

19. Design and Development of Backpack for Carrying Loads by Hill Women

The percentile represents the proportion accommodated for one dimension only. However, while using more than one dimension, the combination of measurements accommodate less than the proportions indicated by the percentiles, therefore, an alternative method to percentiles is the 'Case Selection Method' (Robinette and Hudson, 2004). The purpose of a small number of cases is to simplify the problem by reducing to a minimum amount of information needed. Generally the first thing reduced is the number of dimensions. This is done using knowledge of the product .The goal is to keep just those dimensions that are critical and have as little redundant information as possible. It is easiest if the critical dimensions can be reduced to four or fewer because all the combinations of small and large proportions need to be considered. If the critical dimension is two, then the minimum number of small and large combinations is four, if three then it is eight while more than the four, problem gets quiet complex.

Dimension of the backpack

For the reasons discussed under Section 19 above, only two critical dimensions i.e. length (sitting shoulder height) and the width (shoulder breadth) of the backpack were used and the 90th percentile was used to determine the length and width of the backpack (55 x 39) cm.

Design of the backpack

A total of 100 women from Uttarakhand were selected for calculating the percentiles of the dimensions. The design of the backpack with all the dimensions is shown in the Fig. 4.12 whereas the external frame and first prototype of the backpack are shown in Plates 4.1 to 4.3. The description of the various components of backpack is given below:

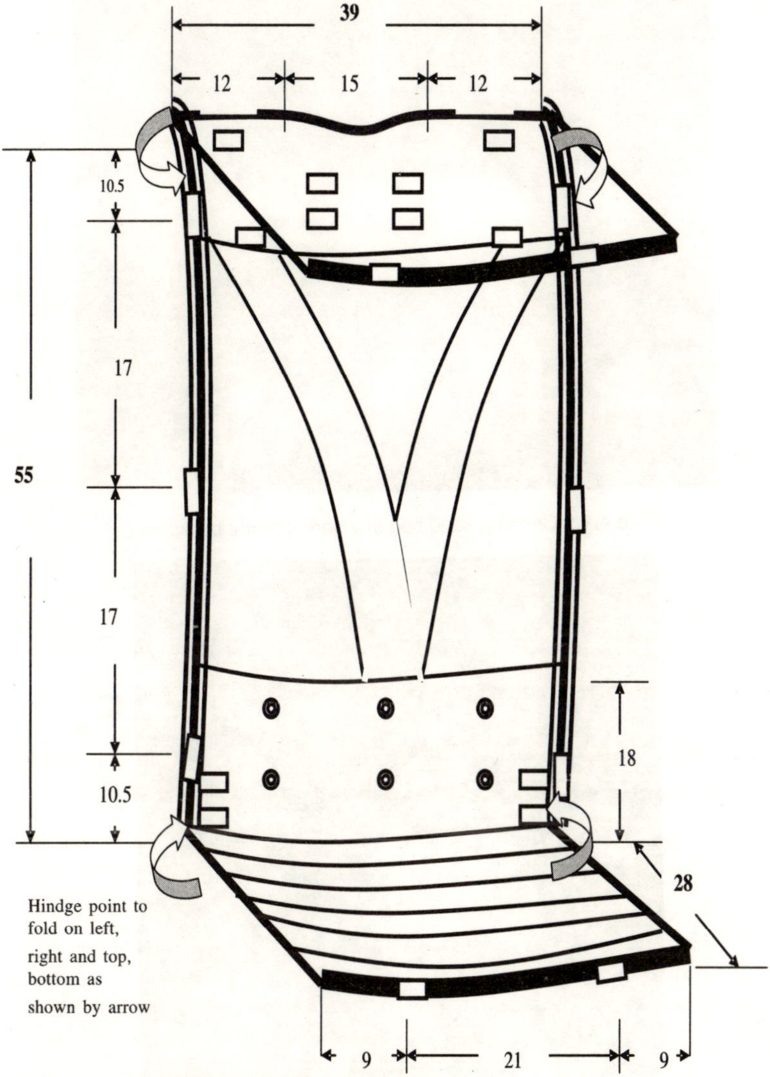

All the measurements are in centimeters,
Material: Flat alluminium alloy.

Holes for straps, Holes for rivets, V-shape – cloth

Fig. 4.12 : External frame of backpack with all the dimensions

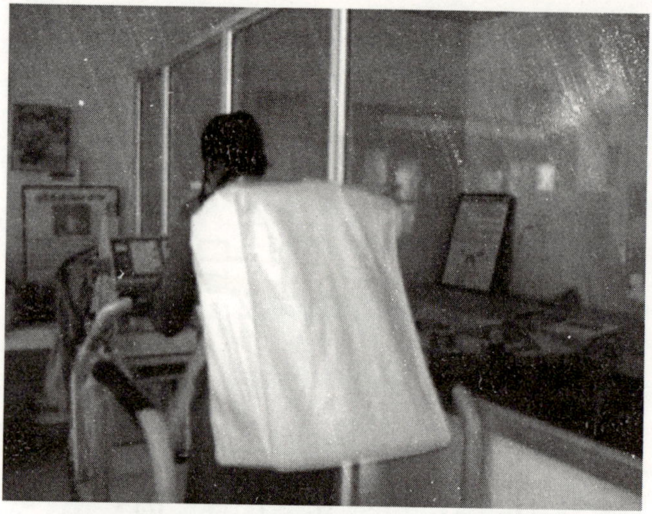

Plate 4.1 : Carrying sand bag in external frame backpack

Plate 4.2 : Carrying grass in external frame backpack

Plate 4.3 : Carrying wood in external frame backpack

(*a*) *External frame:* External frame backpacks are a design where the frame is fully exposed and the bag itself is attached to the sides of the frame. Therefore, a backpack with an external frame was chosen for the design of backpack because of numerous advantages. An external frame generally separates the backpack from the back, thereby providing ample room for air to circulate between the back and backpack. Because of this clearance from the back, external frame backpacks are generally considered to be cooler to carry in warm weather. Also, the vertical bars are slightly curved in S-shape so as to follow the curve of the spinal cord. The shape of the external frame was kept as 'L-shape' so as to support the load from its bottom. The aluminium was used due to its light weight and strength, and the canvas was used for pack profile which was of high quality (10 oz canvas).

(b) *Adjustable shoulder straps:* It was used to distribute the weight uniformly on the whole body.

(c) *Adjustable back and hip belt:* These two belts were used to maintain the balance of the body by keeping the load centered so that the load should not unbalance while walking. The length of the back belt was 84 cm based on the 95th percentile of waist circumference.

(d) *Adjustable chest strap:* They are meant to keep the shoulder straps from sliding off the shoulders under a load. They are not meant to support weight and should never be pulled tightly enough across the chest so as to restrict breathing. The length of the chest belt was 82 cm based on the 95th percentile of chest circumference.

(e) *Adjustable head belt:* This belt was provided as an additional feature based on the fact that most hill women carry a head belt while carrying loads.

The width of all the straps was 3.5 cm

(f) *Hooks or sockets:* These were provided so that the 'L-shape' frame dose not hang while carrying the load.

(g) *Weight of the external frame backpack :* It was approximately 2.0 kg.

(h) *Pack profile:* It is the volume of the backpack. Which was approximately 0.06 m^3 without any load.

Cost calculation

The backpack was fabricated from aluminium frame, heavy duty tape and plastic fitting. Table 4.35 shows material used and cost calculation for external frame backpack.

Table 4.35: Cost calculation for developed backpack

S. No.	Material	Cost, Rs.
1.	Canvas and fitting	200.00
2.	Aluminium frame	600.00
3.	Belts, straps etc.	200.00
4.	Total	1000.00
5.	Taxes @12.5%	125.00
6.	**Total cost**	**1125.00**

20. Performance Evaluation of Developed External Frame Backpack

The performance evaluation of developed backpack was carried out under laboratory and field conditions.

Laboratory condition

Table 4.36 shows the physiological cost while carrying loads with the developed backpack. It was found that the heart rate while carrying loads with developed backpack was 104.7, 110.5 and 114.7 beats/min for 15, 20 and 25 kg loads, respectively whereas for carrying loads on traditional backpack (Plates 3.1 and 3.2) was slightly higher as 106.0, 112.0 and 116.0 beats/min.

Table 4.36 : Physiological cost while carrying loads on backpacks

S. No.	*Loads Weight (kg)*	*Type of back-pack*	*Physiological responses*					
			HR beats/min	*VO_2 l/min*	*EER kcal/min*	*VO_2 /kgml /min/kg*	*METS VO_2 /kg/min*	*RPE Score*
1.	15	D	104.7	0.439	2.21	5.91	1.69	1.00
		T	106.0	0.620	3.11	10.84	3.09	1.85
2.	20	D	110.5	0.561	2.83	7.56	2.16	1.75
		T	112.0	0.710	3.58	12.59	3.59	2.28
3.	25	D	114.7	0.613	3.09	8.25	2.35	2.25
		T	116.0	0.770	3.90	13.34	3.81	3.28

Note: D= Developed externa frame backpack, *T*= Traditional backpack

The OCR values while carrying loads on developed backpack were 0.439, 0.561 and 0.631 l/min with 15, 20 and 25 kg loads, respectively. These values were slightly lower when loads were carried on the traditional backpack as 0.620, 0.710 and 0.770 l/min for 15, 20 and 25 kg loads, respectively.

While carrying loads on traditional backpack, the subjects felt from weak to some what strong exertion whereas with the developed frame backpack the subjects felt very weak exertion.

The values for other physiological parameters viz. EER, VO_2/kg body weight and METS were also found slightly lower for carrying loads with developed backpack than with the traditional backpack, as observed in the case of HR, OCR and RPE.

Field condition

In order to find out the ease of carrying different types of loads in developed backpack, the experiments were also conducted under field conditions. The developed backpack while carrying grass and wood are shown in Plates 4.2 and 4.3, respectively. It was found that the developed backpack was able to accommodate 15 kg of grass and 12.5 kg of the wood with the mean heart rate of 103.5 and 101 beats/min, respectively.

Responses of the subjects on developed backpack

Table 4.37 shows the responses of the subjects regarding the developed backpack. It was found that the subjects were much more comfortable with the developed backpack as far as maintaining the balance, and irritation and pain in shoulder are concerned. The following responses were noticed by the subjects:

- No pinching and pain was felt while using the external frame backpack by the subjects.
- The subjects were able to maintain the natural upright posture, thus it eliminated neck, shoulder and thigh pain to some extent and almost eliminated back pain with heavy

loads as previously felt by the subjects while wearing a traditional backpack (Plates. 3.1 and 3.2) in which case they have to bend forward while carrying loads especially loads carried on 10 per cent grade (G_2).

- Also, subjects did not report any specific pressure points on the body with the developed backpack.

It was, therefore, concluded that the developed L-shaped external frame backpack brought down the level of physiological cost and perceived exertion, and also did not cause any pain and pressure points while carrying loads. However, there is a need for further detailed studies in actual field conditions of hills with different types of loads and materials in order to incorporate improvements, if any in the design to meet the specific requirement.

Table 4.37: Responses of subjects on the developed backpack

S. No.	*Statements*	*Subject's responses*		
		1	2	3
1.	Did you feel any pressure points while carrying the load?	No	No	No
2.	Did you feel any kind of pain due to backpack?	No	No	No
3.	Were you able to maintain the balance of your body while carrying load (sand bag) on the backpack?	Yes*	Yes*	Yes*
4.	Did the chest belt cause problem in breathing?	No	No	No
5.	Did the shoulder straps cause irritation around the shoulder?	No	Yes	No
6.	Was backpack easy to wear?	Yes	Yes	Yes
7.	Is the head belt useful while carrying load?	Yes	Yes	No
8.	Was there any problem due to back and hip belt?	No	No	No
9.	Were you able to maintain the balance of your body while carrying load (grass) on the backpack?	Yes	Yes	Yes
10.	Were you able to maintain the balance of your body while carrying load (wood) on the backpack?	Yes	Yes	Yes
11.	Any other suggestion for improvement.	–	–	–

Note: * for 15 kg only

Conclusion

1. The mean age of the selected hill women of Uttarakhand was 35.86 years. Women in Garhwal region spent more time in collecting fuel, fodder etc. and less time in agricultural activity than in Kumaon region.
2. In Uttarakhand monthly income of the families was very low due to the fact that they had lower productivity from agriculture and livestock.
3. No women from Uttarakhand were found to have any of the deficiency diseases as per their statements.
4. Women felt pain in various body parts such as back, neck, shoulders, calf muscles to name a few and slips and falls during load carrying was very frequent for hill women of Uttarakhand which resulted in injuries in various body parts. Most of Uttarakhand women were having high-average to very good level of physical fitness.
5. Significant differences were not observed in most of the standing and sitting heights yet a few body depths and most of body breadth which showed broader body type of women from Kumaon region probably due to the difference in the geographical conditions and comparatively less physical activity of Kumaon women than Garhwal women because of poor infrastructure facilities in Garhwal region.
6. Selected standing anthropometric measurement of all other states were higher than those of Uttarakhand women except standing height and eye height of Karnataka (Dharwad) women whereas all the selected sitting anthropometric measurements of Andhra Pradesh (Hyderabad), Rajasthan

(Udaipur), Haryana (Hissar) were also higher than the hill women of Uttarakhand.

7. Considering the effect of interaction of three main factors it can be concluded that though the interaction effect was found non-significant, the values were very much higher than the normal working heart rate for different modes of carrying loads at different grades. The highest values for physiological cost and perceived exertion were found while carrying load on shoulder, 25 kg load and on G_2-grade amongst modes, loads and grades, respectively. Whereas lowest was found during carrying 15 kg load on head on H-grade.
8. Effects of all the three interactions were found to be non-significant for the heart rate and increase in the heart rate from resting level. However, interaction effect of load and mode was found to be non-significant with all the physiological parameters except for perceived exertion ($p<0.01$). Also the interaction effect of load and grade was found to be significant ($p<0.01$) for all physiological parameters and RPE except for the heart rate and increase in the heart rate from its resting level. As far as the interaction effect of mode and grade was concerned, it was found to be significant ($p<0.05$) for OCR, EER and METS only.
9. Significant differences were observed regarding main factors i.e. loads, modes and grades in all the physiological parameters and also for perceived exertion. It is concluded that the lowest values of physiological responses and perceived exertion were observed while carrying 15 kg load among loads, carrying on back among different modes and on zero (H) grade among grades.
10. It is concluded that the back mode of carrying load was physiologically least demanding and also perceived to be less demanding than on head and shoulder with all the loads and grades.
11. Heart rate was significantly correlated with the oxygen

consumption rate and rating of perceived exertion and also OCR and RPE were significantly correlated.

12. The time period during which the heart rate and oxygen consumption rate comes to the resting level was considered for recommending the rest period. It was found that the OCR values come to their resting level a little earliar than the heart rate values thus for rest period analysis the heart rate should be used as an indicator.
13. As far as body parts pain and discomfort are concerned, subjects felt light to moderate discomfort and pain in various body parts such as lower back and shoulders while carryings loads on back whereas moderate (with 15 kg load) to very uncomfortable (with 20 and 25 kg load) pain in shoulders, hands, neck and palms while carrying loads on shoulder and felt more than moderate discomfort and pain in head and legs while carrying on head. It was found that the subjects felt moderate discomfort and pain in knees and toes while carrying load on G_2-grade.
14. Different environmental conditions affected the heart rate and perceived exertion. However, in ambient environment the effect on subject's performance might be the result of a combination of different environmental parameters. Thus, there is a need to carry out detailed studies to assess the effect of different environmental conditions (temperature, humidity etc), in controlled environmental chambers, so that the effect of temperature and humidity could be assessed separately and their limits could be decided for comfort.
15. The developed L-shaped external frame backpack resulted in reduced heart rate and RPE values and also it did not result in any kind of pain or pinching in the back thus, it is concluded that the back mode of load carrying has biomechanical and metabolic advantages over other modes of carrying loads.

Implications of the Study

The study brought out a number of important implications given below:

1. *Methodological implications:* Although backpack-mode of carrying has some biomechanical advantages, the hill women should not carry a load of more than 25 kg to avoid a high physiological cost as during carrying 25 kg load especially uphill, the heart rate and OCR increased beyond limits.
2. *Practical implications:* The load, mode and grade have a profound effect on the physiological responses during load carrying hence of special significance for the metabolic rate and should be considered during load carrying activities.

Recommendations for Future Research

1. The similar type of study can be conducted with more subjects and with different types of loads such as grass, grains etc.
2. Studies need to be carried out to assess the effect of environmental conditions (temperature and humidity), in environmentally controlled lab so that the effect of temperature and humidity could be assessed separately and limits for different environmental conditions could be assessed.
3. The developed L-shaped external frame backpack brought down the level of physiological cost and perceived exertion, and also did not cause any pain and pressure points while carrying loads. However, there is a need for further detailed studies in actual field conditions in hills with different types of loads and materials in order to incorporate improvements in the design to meet the specific requirement.

Appendix- A

Table A-1 : ANOVA for anthropometric dimensions of women of Kumaon and Garhwal regions

S. No.	*Source*	*d.f.*	*S.S*	*M.S*	*F-value*
1.	**Body Weight**				
	Treat.	1	0.2808 E+02	0.2808 E+02	1.086 ns
	Error	98	0.2534 E+04	0.2585 E+02	
	Total	99	0.2562 E+04		
	C.D. at 5 per cent		0.201809 E+01 ns	C.D. at 1 per cent	0.267157 E+01 ns
2.	**Standing Height**				
	Treat.	1	0.2500E+00	0.2500 E+00	0.007 ns
	Error	98	0.3402 E+02	0.3472 E+02	
	Total	99	0.3403 E+04		
	C.D. at 5 per cent		0.233867 E+01 ns	C.D. at 1 per cent	0.309595 E+01 ns
3.	**Eye Height Standing**				
	Treat.	1	0.2466 E+03	0.2466 E+03	5.101*
	Error	98	0.4738 E+04	0.4835 E+02	
	Total	99	0.4985 E+04		
	C.D. at 5 per cent		0.275973 E+01 *	C.D. at 1 per cent	0.365336 E+01ns
4.	**Shoulder Height Standing**				
	Treat.	1	0.1600 E+02	0.1600 E+02	0.608 ns
	Error	98	0.2580 E+04	0.2633 E+02	
	Total	99	0.2596 E+04		
	C.D. at 5 per cent		0.203647 E+01 ns	C.D. at 1 per cent	0.269591E+01ns
5.	**Elbow Height Standing**				
	Treat.	1	0.7056 E+02	0.7056 E+02	2.0618 ns
	Error	98	0.2641 E+04	0.2695 E+02	
	Total	99	0.2711 E+04		
	C.D. at 5 per cent		0.206036 E+01 ns	C.D. at 1 per cent	0.272752 E+01ns
6.	**Knuckle Height Standing**				
	Treat.	1	0.1688 E+01	0.1688 E+01	0.034 ns
	Error	98	0.4874 E+04	0.4974 E+02	
	Total	99	0.4876 E+04		
	C.D. at 5 per cent		0.279920 E+01 ns	C.D. at 1 per cent	0.370561 E+01ns
7.	**Over head Reach**				
	Treat.	1	0.6075 E+02	0.6075 E+02	1.0521 ns
	Error	98	0.3914 E+04	0.3994 E+02	
	Total	99	0.3975 E+04		
	C.D. at 5 per cent		0.250838 E+01 ns	C.D. at 1 per cent	0.332062 E+01ns

S. No.	Source	d.f.	S.S	M.S	F-value
8.	**Sitting Height**				
	Treat.	1	0.6394 E+02	0.6394 E+02	4.880*
	Error	98	0.1284 E+04	0.1310 E+02	
	Total	99	0.1348 E+04		
	C.D. at 5 per cent		0.143658 E+01 *	C.D. at 1 per cent	0.190176 E+01 ns
9.	**Eye height Sitting**				
	Treat.	1	0.4356 E+02	0.4356 E+02	2.931 ns
	Error	98	0.1456 E+04	0.1486 E+02	
	Total	99	0.1500 E+04		
	C.D. at 5 per cent		0.153008 E+01 ns	C.D. at 1 per cent	0.202554 E+01 ns
10.	**Shoulder Height Sitting**				
	Treat.	1	0.3970 E+02	0.3970 E+02	2.947 ns
	Error	98	0.1320 E+04	0.1347 E+02	
	Total	99	0.1360 E+04		
	C.D. at 5 per cent		0.145670 E+01 ns	C.D. at 1 per cent	0.192840E+01 ns
11.	**Elbow Height Sitting**				
	Treat.	1	0.1850 E+03	0.1850 E+03	31.862**
	Error	98	0.5689 E+03	0.5805 E+01	
	Total	99	0.7538 E+03		
	C.D. at 5 per cent		0.956270 E+00 *	C.D. at 1 per cent	0.126592 E+01 **
12.	**Thigh Height sitting**				
	Treat.	1	0.8702 E+01	0.8702 E+01	3.016 ns
	Error	98	0.2827 E+03	0.2885 E+01	
	Total	99	0.291403		
	C.D. at 5 per cent		0.674165 E+00 ns	C.D. at 1 per cent	0.892467 E+00
13.	**Knee Height Sitting**				
	Treat.	1	0.4623 E+02	0.4623 E+02	2.164 ns
	Error	98	0.2094 E+04	0.2136 E+02	
	Total	99	0.2140 E+04		
	C.D. at 5 per cent		0.183454 E+01 ns	C.D. at 1 per cent	0.242859 E+01 ns
14.	**Poplietal Height Sitting**				
	Treat.	1	0.1953 E+01	0.1953 E+01	0.129 ns
	Error	98	0.1486 E+04	0.1517 E+02	
	Total	99	0.1488 E+04		
	C.D. at 5 per cent		0.154564 E+01 ns	C.D. at 1 per cent	0.204613 E+01 ns
15.	**Buttock-Leg Length**				
	Treat.	1	0.6888 E+02	0.6888 E+02	2.011 ns
	Error	98	0.3356 E+04	0.3424 E+02	
	Total	99	0.3424 E+04		
	C.D. at 5 per cent		0.232250 E+01 ns	C.D. at 1 per cent	0.307455 E+01 ns
16.	**Buttock-knee Length**				
	Treat.	1	0.1089 E+02	0.1089 E+02	0.642 ns
	Error	98	0.1662 E+04	0.1696 E+02	
	Total	99	0.1673 E+04		
	C.D. at 5 per cent		0.163443 E+01 ns	C.D. at 1 per cent	0.216367 E+01 ns
17.	**Buttock-Poplietal Length**				
	Treat.	1	0.1295 E+02	0.1295 E+02	0.827 ns
	Error	98	0.1535 E+04	0.1567 E+02	
	Total	99	0.154804		
	C.D. at 5 per cent		0.157091 E+01 ns	C.D. at 1 per cent	0.207959 E+01 ns
18.	**Hand Breadth at Metacarpal**				
	Treat.	1	0.1025E-01	0.1025E-01	0.036 ns
	Error	98	0.2798 E+02	0.2855 E+01	

S. No.	Source	d.f.	S.S	M.S	F-value
	Total	99	0.2798 E+02		
	C.D. at 5 per cent		0.212077 E+00ns	C.D. at 1 per cent	0.280749 E+00 ns
19.	**Hand Breadth at Thumb**				
	Treat.	1	0.4902 E+00	0.4902 E+00	1.194 ns
	Error	98	0.4025 E+02	0.4107 E+00	
	Total	99	0.4074 E+02		
	C.D. at 5 per cent		0.254362 E+00 ns	C.D. at 1 per cent	0.336727 E+00 ns
20.	**Hand Length**				
	Treat.	1	0.2102 E+01	0.2102 E+01	2.103 ns
	Error	98	0.9795 E+02	0.9994 E+00	
	Total	99	0.1000 E+03		
	C.D. at 5 per cent		0.396790 E+00 ns	C.D. at 1 per cent	0.525275 E+00 ns
21.	**Hand Thickness**				
	Treat.	1	0.1156 E+01	0.1156 E+01	31.831**
	Error	98	0.35558 E+01	0.3631 E-01	
	Total	99	0.4714 E+01		
	C.D. at 5 per cent		0.756250 E-01*	C.D. at 1 per cent	0.100113 E+00 **
22.	**Shoulder-Elbow Length**				
	Treat.	1	0.3480 E+02	0.3480 E+02	6.881*
	Error	98	0.4957 E+03	0.5058 E+01	
	Total	99	0.5305 E+03		
	C.D. at 5 per cent		0.892647 E+00*	C.D. at 1 per cent	0.118170 E+01 ns
23.	**Forearm-Hand Length**				
	Treat.	1	0.8109 E+01	0.8109 E+01	1.242 ns
	Error	98	0.6398 E+03	0.6529 E+01	
	Total	99	0.6479 E+03		
	C.D. at 5 per cent		0.101413 E+01 ns	C.D. at 1 per cent	0.134252 E+01 ns
24.	**Arm Reach**				
	Treat.	1	0.1769 E+02	0.1769 E+02	1.423 ns
	Error	98	0.1218 E+04	0.1243 E+02	
	Total	99	0.1236 E+04		
	C.D. at 5 per cent		0.139917 E+01 ns	C.D. at 1 per cent	0.185244 E+01 ns
25.	**Interpupillary Distance**				
	Treat.	1	0.4492 E+00	0.4492 E+00	0.414 ns
	Error	98	0.1063 E+03	0.1084 E+01	
	Total	99	0.1067 E+03		
	C.D. at 5 per cent		0.413320 E+00 ns	C.D. at 1 per cent	0.547157 E+00 ns
26.	**Head Length**				
	Treat.	1	0.3238 E+01	0.3238 E+01	1.082 ns
	Error	98	0.2933 E+03	0.2993 E+01	
	Total	99	0.2966 E+03		
	C.D. at 5 per cent		0.686657 E+00 ns	C.D. at 1 per cent	0.909004 E+00 ns
27.	**Head Breadth**				
	Treat.	1	0.8464 E+02	0.8464 E+02	23.463**
	Error	98	0.3535 E+03	0.3607 E+01	
	Total	99	0.4382 E+03		
	C.D. at 5 per cent		0.753834 E+00*	C.D. at 1 per cent	0.997933 E+00**
28.	**Shoulder Breadth**				
	Treat.	1	0.1850 E+03	0.1850 E+03	30.156**
	Error	98	0.6011 E+03	0.6133 E+01	
	Total	99	0.7860 E+03		
	C.D. at 5 per cent		0.982957 E+00*	C.D. at 1 per cent	0.130125 E+01**

S. No.	Source	d.f.	S.S	M.S	F-value
29.	**Chest Breadth**				
	Treat.	1	0.1600 E+02	0.1600 E+02	3.086 ns
	Error	98	0.5082 E+03	0.5185 E+01	
	Total	99	0.5242 E+03		
	C.D. at 5 per cent		0.903790 E+00 ns	C.D. at 1 per cent	0.119645 E+01 ns
30.	**Maximum Body Breadth**				
	Treat.	1	0.2862 E+02	0.2862 E+02	1.710 ns
	Error	98	0.1640 E+04	0.1674 E+02	
	Total	99	0.1669 E+04		
	C.D. at 5 per cent		0.162371 E+01 ns	C.D. at 1 per cent	0.214948 E+01 ns
31.	**Elbow-Elbow Breadth**				
	Treat.	1	0.4995 E+03	0.4995 E+03	29.461**
	Error	98	0.1662 E+04	0.1696 E+02	
	Total	99	0.2161 E+04		
	C.D. at 5 per cent		0.163431 E+01*	C.D. at 1 per cent	0.216352 E+01**
32.	**Hip Breadth sitting**				
	Treat.	1	0.7921 E+02	0.7921 E+02	14.697**
	Error	98	0.5282 E+03	0.5390 E+01	
	Total	99	0.6074 E+03		
	C.D. at 5 per cent		0.921425 E+00*	C.D. at 1 per cent	0.121979 E+01**
33.	**Hip Breadth Standing**				
	Treat.	1	0.8414 E+01	0.8414 E+01	2.169 ns
	Error	98	0.3802 E+03	0.3879 E+01	
	Total	99	0.3886 E+03		
	C.D. at 5 per cent		0.78174200 E+00 ns	C.D. at 1 per cent	0.103488 E+01 ns
34.	**Knee-Knee Breadth**				
	Treat.	1	0.1145 E+03	0.1145 E+03	41.85**
	Error	98	0.2681 E+03	0.2736 E+01	
	Total	99	0.3826 E+03		
	C.D. at 5 per cent		0.656473 E+00*	C.D. at 1 per cent	0.869046 E+00**
35.	**Chest Circumference**				
	Treat.	1	0.1010 E+03	0.1010 E+03	2.418 ns
	Error	98	0.4094 E+04	0.4178 E+02	
	Total	99	0.4195 E+04		
	C.D. at 5 per cent		0.256537 E+01 ns	C.D. at 1 per cent	0.339606 E+01 ns
36.	**Waist Circumference**				
	Treat.	1	0.8940 E+03	0.8940 E+03	19.095 **
	Error	98	0.4588 E+04	0.4682 E+02	
	Total	99	0.5482 E+04		
	C.D. at 5 per cent		0.271575 E+01*	C.D. at 1 per cent	0.359514 E+01**
37.	**Chest Depth**				
	Treat.	1	0.9604 E+02	0.9604 E+02	11.292**
	Error	98	0.9335 E+03	0.8505 E+01	
	Total	99	0.9295 E+03		
	C.D. at 5 per cent		0.115747 E+01*	C.D. at 1 per cent	0.153228 E+01**
38.	**Waist Depth**				
	Treat.	1	0.3721 E+02	0.3721 E+01	6.396 *
	Error	98	0.5701 E+03	0.5817 E+01	
	Total	99	0.6073 E+03		
	C.D. at 5 per cent		0.957293 E+00*	C.D. at 1 per cent	0.126728 E+01 ns
39.	**Maximum Body Depth**				
	Treat.	1	0.2372 E+03	0.2372 E+03	26.203 **
	Error	98	0.8870 E+03	0.9051 E+01	

S. No.	Source	d.f.	S.S	M.S	F-value
	Total	99	0.1124 E+04		
	C.D. at 5 per cent		0.119407 E+01*	C.D. at 1 per cent	0.158073 E+01 **
40.	**Buttock Depth**				
	Treat.	1	0.6406 E+00	0.6406 E+00	0.181 ns
	Error	98	0.3475 E+03	0.3546 E+01	
	Total	99	0.3482 E+03		
	C.D. at 5 per cent		0.747409 E+00 ns	C.D. at 1 per cent	0.989428 E+00 ns
41.	**Foot Breadth**				
	Treat.	1	0.1156 E+02	0.1156 E+02	13.537**
	Error	98	0.8369 E+02	0.8540 E+00	
	Total	99	0.9525 E+02		
	C.D. at 5 per cent		0.366779 E+00*	C.D. at 1 per cent	0.485546 E+00**
42.	**Foot Length**				
	Treat.	1	0.3906 E-01	0.3906E-01	0.016 ns
	Error	98	0.2323 E+03	0.2370 E+01	
	Total	99	0.2323 E+03		
	C.D. at 5 per cent		0.611075 E+00 ns	C.D. at 1 per cent	0.808947 E+00 ns

Appendix - B

ANOVA for physiological parameters and perceived exertion

Table B-1: Analysis of variance for mean heart rate

S. No.	Source of variation	d.f.	S.S	M.S	F-value	
1.	Load	2	5579.238	2789.619	55.60092	**
2.	Mode	2	4937.207	2468.603	49.20264	**
3.	Grade	2	11311.30	5655.651	112.7248	**
4.	Load * Mode	4	51.17422	12.79356	0.2549931	ns
5.	Mode * Grade	4	86.15836	21.53959	0.4293135	ns
6.	Load * Grade	4	107.1742	26.79356	0.5340323	ns
7.	Load * Mode * Grade	8	50.60324	6.325405	0.1260740	ns
8.	Error	162	8127.893	50.17218		
9.	Total	188	30250.75	160.9082		

SEm± i,j,k,ij,ik,jk,ijk : 0.8924034 0.8924034 0.8924034 1.5456879 1.5456879 1.5456879 2.6772101

C. D. at 5 per cent : 2.4922342 2.4922342 2.4922342 4.3166761 4.3166761 4.3166761 7.4767027

C.V. = 5.679881

Table B-2: Analysis of variance for increase in mean heart rate

S. No.	Source of variation	d.f.	S.S	M.S	F-value	
1.	Load	2.	5579.238	2789.619	55.60092	**
2.	Mode	2.	4937.207	2468.603	49.20264	**
3.	Grade	2.	11311.30	5655.651	112.7248	**
4.	Load * Mode	4.	51.17422	12.79356	0.2549931	ns
5.	Mode * Grade	4.	86.15836	21.53959	0.4293135	ns
6.	Load * Grade	4.	107.1742	26.79356	0.5340323	ns
7.	Load * Mode * Grade	8.	50.60324	6.325405	0.1260740	ns
8.	Error	162.	8127.893	50.17218		
9.	Total	188.	30250.75	160.9082		

SEm± i,j,k,ij,ik,jk,ijk : 0.8924034 0.8924034 0.8924034 1.5456879 1.5456879 1.5456879 2.6772101

C.D. at 5 per cent : 2.4922342 2.4922342 2.4922342 4.3166761 4.3166761 4.3166761 7.4767027

C.V. = 5.679881

Table B-3 : Analysis of variance for OCR

S.No.	Source of variation	d.f.	S.S	M.S	F-value	
1.	Load	2	1.333063	0.6665317	141.6509	**
2.	Mode	2	1.968573	0.9842866	209.1800	**
3.	Grade	2	3.397501	1.698751	361.0175	**
4.	Load * Mode	4	0.4185899E-01	0.1046475E-01	2.223962	ns
5.	Mode * Grade	4	0.5101220E-01	0.1275305E-01	2.710271	*
6.	Load * Grade	4	0.5146566E-01	0.1286642E-01	2.734363	**
7.	Load * Mode*Grade	8	0.3155565E-01	0.3944457E-02	0.8382735	ns
8.	Error	162	0.7622833	0.4705453E-02		
9.	Total	188	7.637314	0.4062401E-01		

SEm± i,j,k,ij,ik,jk,ijk : 0.0086423 0.0086423 0.0086423 0.0149689 0.0149689 0.0149689 0.0259270

C.D. at 5 per cent : 0.0241356 0.0241356 0.0241356 0.0418041 0.0418041 0.0418041 0.0724068

C.V. = 7.237915

Table B-4: Analysis of variance for EER

S.No.	Source of variation	df	S.S	M.S	F-value	
1.	Load	2	33.95547	16.97774	142.8103	**
2.	Mode	2	49.95448	24.97724	210.0991	**
3.	Grade	2	87.11470	43.55735	366.3880	**
4.	Load * Mode	4	1.060024	0.2650061	2.229131	ns
5.	Mode * Grade	4	1.379594	0.3448984	2.901155	*
6.	Load * Grade	4	1.253974	40.3134935	2.636989	**
7.	Load * Mode * Grade	8	0.8932909	0.1116614	0.9392532	ns
8.	Error	162	19.25907	0.1188831		
9.	Total	188	194.8706	1.036546		

SEm± i,j,k,ij,ik,jk,ıjk : 0.0434400 0.0434400 0.0434400 0.0752403 0.0752403 0.0752403 0.1303200

C.D. at 5 per cent : 0.1213158 0.1213158 0.1213158 0.2101252 0.2101252 0.2101252 0.3639475**C.V.** = 7.209590

Table B-5: Analysis of variance for OCR/kg body weight

S.No.	Source of variation	d.f.	S.S	M.S	F-value	
1.	Load	2	411.4090	205.7045	111.2632	**
2.	Mode	2	628.5082	314.2541	169.9764	**
3.	Grade	2	1066.052	533.0259	288.3075	**
4.	Load * Mode	4	14.34296	3.585741	1.939486	ns
5.	Mode * Grade	4	17.79535	4.448837	2.406324	ns
6.	Load * Grade	4	13.33504	3.333760	1.803192	**
7.	Load * Mode * Grade	8	11.60505	1.450631	0.7846296	ns
8.	Error	162	299.5073	1.848810		
9.	Total	188	2462.555	13.09869		

SEm± i,j,k,ij,ik,jk,ijk : 0.1713073 0.1713073 0.1713073 0.2967130 0.2967130 0.2967130 0.5139219

C.D at 5 per cent : 0.4784136 0.4784136 0.4784136 0.8286367 0.8286367 0.8286367 1.4352409

C.V. = 8.112339

Table B-6: Analysis of variance for METS

S.No.	*Source of variation*	*d.f.*	*S.S*	*M.S*	*F-value*	
1.	Load	2	33.58389	16.79195	111.3498	**
2.	Mode	2	51.30562	25.65281	170.1075	**
3.	Grade	2	87.01296	43.50648	288.4977	**
4.	Load * Mode	4	1.167967	0.2919917	1.936239	ns
5.	Mode * Grade	4	1.455043	0.3637607	2.412149	*
6.	Load * Grade	4	1.080786	0.2701966	1.791712	**
7.	Load * Mode *Grade	8	0.9503784	0.1187973	0.7877620	ns
8.	Error	162	24.43018	0.1508036		
9.	Total	188	200.9868	1.069079		

SEm± i,j,k,ij,ik,jk,ijk: 0.0489255 0.0489255 0.0489255 0.0847415 0.0847415 0.0847415 0.1467766

C.D. at 5 per cent: 0.1366354 0.1366354 0.1366354 0.2366594 0.2366594 0.2366594 0.4099061

C.V. = 8.109696

Table B-7: Analysis of variance for RPE

S.No.	*Source of variation*	*d.f.*	*S.S*	*M.S*	*F-value*	
1.	Load	2	176.0423	88.02116	271.2391	**
2.	Mode	2	370.4550	185.2275	570.7826	**
3.	Grade	2	101.3439	50.67196	156.1467	**
4.	Load * Mode	4	6.052925	1.513231	4.663055	**
5.	Mode * Grade	4	0.8465729	0.2116432	0.6521832	ns
6.	Load * Grade	4	11.06879	2.767196	8.527176	**
7.	Load * Mode * Grade	8	8.264509	1.033064	3.183408	**
8.	Error	162	52.57143	0.3245150		
9.	Total	188	726.6455	3.865136		

SEm± i,j,k,ij,ik,jk,ijk: 0.0717707 0.0717707 0.0717707 0.1243105 0.1243105 0.1243105 0.2153121

C.D. at 5 per cent: 0.2004356 0.2004356 0.2004356 0.3471646 0.3471646 0.3471646 0.6013067

C.V. = 11.20355

Appendix - C

Table C-1: Regression analysis between OCR and HR

0.9477778 y - bar

1 0.1729042E-01 (b) 124.7933 (x-bar)

0.9423537 due to reg. -1.209951 (a) 0.1460517E-02 err.mss 0.9788666 total ss 0.962699 r-sq f-value = 645.2190 1 0.6806938E-03 s.e. 25.40116 t-value

Table C-2: Regression analysis between HR and OCR

124.7933 y - bar

1 55.67817 (b) 0.9477778 (x-bar)

3034.544 due to reg. 72.02280 (a) 4.703127 err.mss 3152.122 total ss 0.962699 r-sq f-value = 645.2179 1 2.191955 s.e. 25.40115 t-value

Table C-3: Regression analysis between RPE and HR

5.080370 y - bar

1 0.1469270 (b) 124.7933 (x-bar)

68.04659 due to reg. -13.25514 (a) 1.131252 err.mss 96.32790 total ss 0.706406 r-sq f-value = 60.15155 1 0.1894429E-01 s.e. 7.755743 t-value

Table C-4: Regression analysis between HR and RPE

124.7933 y - bar

1 4.807869 (b) 5.080370 (x-bar)

2226.678 due to reg. 100.3676 (a) 37.01779 err.mss 3152.122 total ss 0.706406 r-sq f-value = 60.15157 1 0.6199108 s.e. 7.755744 t-value

Table C-5: Regression analysis between RPE and OCR

5.080370 y - bar

1 8.656871 (b) 0.9477778 (x-bar)

73.35765 due to reg. -3.124420 (a) 0.9188099 err.mss 96.32790 total ss 0.761541 r-sq f-value = 79.83986 1 0.9688378 s.e. 8.935315 t-value

Table C-6: Regression analysis between OCR and RPE

0.9477778 y - bar

1 0.8796955E-01 (b) 5.080370 (x-bar)

0.7454471due to reg. 0.5008599 (a) 0.9336780E-02 err.mss 0.9788666 total ss 0.761541 r-sq f-value = 79.83986 1 0.9845154E-02 s.e. 8.935315 t-value

Appendix - D

Table D-1: Regression analysis between heart rate and independent variables (grade, mode and load)

124.7933 y - bar

1 9.527224 (b) 2.000000 (x-bar)

2 6.098334 (b) 2.000000 (x-bar)

3 1.286444 (b) 20.00000 (x-bar)

3047.961 due to reg. 67.81333 (a) 4.528768 err.mss 3152.122 total ss

.966955 r-sq

f-value = 224.3407

1 .5015957 s.e. 18.99383 t-value

2 .5015957 s.e. 12.15787 t-value

3 .1003191 s.e. 12.82352 t-value

Table D-2: Regression analysis between OCR and independent variables (grade, mode and load)

.9477778 y - bar

1.1627778 (b) 2.000000 (x-bar)

2.1194444 (b) 2.000000 (x-bar)

3.2177778E-01 (b) 20.00000 (x-bar)

.9471666 due to reg. -.5222218E-01 (a).1378262E-02 err.mss .9788666 total ss

.967616 r-sq

f-value = 229.0724

1 .8750436E-02 s.e. 18.60225 t-value

2 .8750436E-02 s.e. 13.65011 t-value

3 .1750087E-02 s.e. 12.44382 t-value

Appendix - F

INTERVIEW SCHEDULE

Schedule number————————————

1. Name of the respondent————————————

2. **Demographic profile**

S.No.	*Demographic characteristics*	
1.	Mean age (Years)	
2.	Education	
	(i) Illiterate	
	(ii) Primary	
	(iii) Junior high school	
	(iv) High school	
	(v) Intermediate	
	(vi) Graduate	
	(vii) Post-graduate	
3.	Family type	
	(i) Nuclear	
	(ii) Joint	
	(iii) Extended	
4.	Family size	
	(i) 1-3 (Small)	
	(ii) 4-7 (Medium)	
	(iii) > 8 (Large)	
5.	Source of income	
	(i) Agriculture	
	(ii) Business	
	(iii) Service	
	(iv) Livestock (Dairy)	
6.	Monthly income (Rs)	
	(i) 500-1000	

(ii) 1001-2000	
(iii) 2001-3000	
(iv) 3001-4000	
(v) 4001-5000	
(vi) >5000	
7. Average time spent in (h/day)	
(i) Collecting fuel, fodder etc	
(ii) Agricultural activity	

3. Medical history in last six months

S.No *Illness*	
1. Common Illness	
(i) Cold, cough	
(ii) Headache, fever	
(iii) Digestive problems	
(iv) Hearing and sight problems	
2. Chronic Illness	
(i) Diabetes	
(ii) B.P	
(iii) Respiratory	
3. Deficiency diseases	
(i) Fe-deficiency	
(ii) Ca-deficiency	
(iii) Any other	

4. Pain in various body parts of Uttarakhand women during carrying loads in last six months

S.No. *Body parts*	
(i) Body pain	
(ii) Shoulder pain	
(iii) Neck pain	
(iv) Back pain	
(v) Forearm pain	
(vi) Pain in palms	
(vii) Knee pain	
(viii) Foot pain	

(ix) Pain in calf muscles (x) Pain in biceps (xi) Slips (xii) Falls (xiii) Cuts and wounds, injuries (xiv) Fractures (xv) Swelling	

5. Physical fitness index table

S. No.	*Heart rate in Recovery Period*			*PFI score*
	1st min	2nd min	3rd min	

6. Anthropometric measurement

S.No.	*Anthropometric measurement*	*Value*
1.	Weight (kg)	
2.	Height	
3.	Eye height	
4.	Shoulder height	
5.	Elbow height standing	
6.	Knuckle height standing	
7.	Overhead reach	
8.	Sitting height	
9.	Eye height	
10.	Shoulder height	
11.	Elbow height	
12.	Thigh height	
13.	Knee height	
14.	Poplietal height	
15.	Buttock-leg length	
16.	Buttock-knee length	
17.	Buttock-poplietal length	
18.	Hand breadth at metacarpal	
19.	Hand breadth at thumb	
20.	Hand length	

21. Hand thickness
22. Shoulder-elbow length
23. Forearm-hand length
24. Arm reach
25. Interpupillary distance
26. Head length
27. Head breadth
28. Shoulder breadth
29. Chest breadth
30. Maximum body breadth
31. Elbow-elbow breadth
32. Hip breadth sitting
33. Hip breadth standing
34. Knee-Knee breadth
35. Chest circumference
36. Waist circumference
37. Chest depth
38. Waist depth
39. Maximum body depth
40. Buttock depth
41. Foot breadth
42. Foot length

7. Physiological cost in terms of ....

Level of Independent variables				*Physiological cost*					*Recovery period*
				Time					
S. No.	*Load*	*Mode*	*Grade*	*5th*	*6th*	*7th*	*8th*	*Mean*	
1.	15	Back	H						
2.	15	Back	G_1						
3.	15	Back	G_2						
4.	15	Head	H						
5.	15	Head	G_1						
6.	15	Head	G_2						
7.	15	Shoulder	H						
8.	15	Shoulder	G_1						
9.	15	Shoulder	G_2						
10.	20	Back	H						
11.	20	Back	G_1						

12.	20	Back	G_2
13.	20	Head	H
14.	20	Head	G_1
15.	20	Head	G_2
16.	20	Shoulder	H
17.	20	Shoulder	G_1
18.	20	Shoulder	G_2
19.	25	Back	H
20.	25	Back	G_1
21.	25	Back	G_2
22.	25	Head	H
23.	25	Head	G_1
24.	25	Head	G_2
25.	25	Shoulder	H
26.	25	Shoulder	G_1
27.	25	Shoulder	G_2

Bibliography

Abe, D.; Yanagawa K. and Niihata, S. 2004. Effects of load carriage, load position, and walking speed on energy cost of walking. *Applied Ergonomics,* 35 (4):229-235.

Adak, Dipak Kumar. 2006. Body mass index and chronic energy deficiency of adult males of central Indian populations *Human Biology,* 78 (2): 161-178.

Balogun, J.A. 1986. Ergonomic comparison of three modes of load carriage. *Int. Arch. Occup. Environ. Health,* 58, 35-46.

Bastien, G. J.; Willems, P. A.; Schepens, B. and Heglund N. C. 2005. Effect of load and speed on the energetic cost of human walking. *European Journal of Applied Physiology,* 94 (1-2): 76-83.

Borg, G. 1982. Psychophysical bases of perceived exertion. *Medical Science in Sports and Exercise,* 14:377-38.

Borg, G. A.V. 1970. Perceived exertion as an indicator of somatic stress. *Scandinavian Journal of Rehabilitative Medicine,* 2:92-98.

Borghols, E.A.; Dresen, M.H. and Hollander, A.P. 1978. Influence of heavy weight carrying on the cardio respiratory system during exercise. *Eur J Appl Physiol Occup Physiol.,* 38 (3):161-169.

Bose, A. 2002. Missing men and lonely women: Demography of Himalayan villages. *Economic and Political Weekly.* 27 (July): 1-7.

Chakrabarti, Debkumar, 1997. Indian anthropometric dimensions for ergonomic design practice. National Institute of Design, Ahemdabad, India.

Chand, Raja and Singh, Raghbir. 1978. Estimation of body fat and lean body mass from anthropometric dimensions in adult Indian women. *Annals,* 5 (6): 591-594.

Charteris, J.; Nottrodt, J.W. and Scott, P.A. 1989. The "free-ride" hypothesis: A second look at the efficiency of African women head load carriers. *S. Afr. J. Sci.,* (85): 68-71.

Charteris, J.; Scott, P.A. and Nottrodt, J.W. 1989. Metabolic and kinematic responses of African women head load carriers under controlled conditions of load and speed. *Ergonomics,* (32) :1539-1550.

Christensen, E. H. 1953. Physiological variations in work in Nykroppa Iron Works. Ergonomics Society Symposium on Fatigue. London. Lewis, pp. 93-108.

Christensen, H.; Pedersen, M.B. and Sjogaard, G. 1995. A national cross sectional study in the Danish wood and furniture industry on working postures and manual material handling. *Ergonomics,* 38 (4):793-805.

Chung, M.K.; Lee, V.J.; Lee, I. and Choi, K.I. 2005. Physiological workload evaluation of carrying soft drink beverage boxes on the back. *Journal of Applied Ergonomics,* 36 (5):569-574.

Corlett, E. N. and Bishop, R. P. 1976. A technique for assessing postural discomfort. *Ergonomics,* 19 (2): 175-182.

Daijiro, Abe.; Kazumassa, Y. and Shigemitsu, N. 2004. Effects of load carriage, load position and walking speeds on energy cost of walking. *Applied Ergonomics,* 35:329-335.

Datta, S.R. and Ramanathan, N.L. 1970. Ergonomic comparisons of seven modes of carrying loads on the horizontal plane. *Ergonomics,* 14: 269-278.

De Zwart, B.C.H.; Frings-Dresen, M.H.W. and Kilbom, A. 2001. Gender differences in upper extremity muskuloskeletal complaints in the working population. *Int. Arch Occup. Environ. Health.,* 74 : 21-30.

Dewangan, K. N.; Prasanakumar, G.V.; Sujap, L.; and Choudhary, M.D. 2005. Anthropometric dimensions of farm youth of the north eastern region of India. *International Journal of Industrial Ergonomics,* (11): 979-989.

Ducharme, R.E. 1973. Problems tools for women Indust. Eng. Syst., pp. 46-50.

Eschenbacher, W.L., Moore, T.B., Lorenzen T.J., Weg, J.G., and Gross, K.B., 1992. Pulmonary responses of asthmatic and normal subjects to different temperature and humidity conditions in an environmental chamber. *Lung,* 170 (1):51-62.

Falola, J.M., Delpech, N., and Brisswalter, J.2000. Optimization characteristics of walking with and without a load on the trunk of the body. *Percept MotSkills,* 91 (1):261-272.

Francis, K and Hobbler, T., 1985. Changes in oxygen consumption associated with treadmill walking and running with light hand-carried weight. *Ergonomics,* 35:417-426.

Gite, L. P. and Yadav, B. G. 1989. Anthropometric survey for agricultural machinery design-An Indian case study. *Applied Ergonomics,*:191-196.

Gite, L.P., and Gingh, G., 1997. Ergonomics in agricultural and allied activities in India .Technical Bulletin No. CIAE/(97/ 70).Central Institute of Agricultural Engineering, Bhopal, India.

Gleim, G.W., and Nicholas, J.A., (1989). Metabolic costs and heart rate responses to treadmill walking in water at different depths and temperatures. *Am. J. Sports Med.,* 17 (2):248-252.

Gordon, M.J., Goslin, B.R., Graham, T., and Hoare, J., 1983. Comparisons between load carriage and grade walking on a treadmill. *Ergonomics,* 26:289-298.

Grandjean, E., 1982. Fitting the Task to the Man—An Ergonomic Approach. Taylor and Francis Ltd, London. pp 77-81.

Gupta, P. K., Gupta, M.L. and Sharma, A.P. 1983. Anthropometric survey of Indian farm workers. *Agric. Mech. Asia, Africa and Latin America,* 14 (1):27-30.

Gupta, S. J., Dimri, G. P. and Malhotra, M. S. 1977. Metabolic responses of Indians during sub-maximal work in dry and humid heat. *Ergonomics,* 20 (1):27-30.

Heglund, N.C., Willems, P.A., Penta, M., and Cavagna, G.A., 1995. Energy-saving gait machines with head–supported loads. *Nature,* 375:52-54.

Hooper, T.L., Dunn, D.M., Props, J.E., Sawyer, S.F and Daniel, J.A., 2004. The effects of graded forward and backward walking on heart rate and oxygen consumption. *J. Orthop. Sports Phys. Ther.,* 34 (2):65-71.

Hsiao, H. 2002. Anthropometric differences among occupational groups. *Ergonomics,* 45 (2):136-151.

Ishida, M. 1975. The physiological responses to walking uphill under hot environment. *Bull Tokyo Med. Dent. Univ.,* 22 (2):127-44.

James, N.; Darlene, A.S. and Park, K.S. 2004. Physiological responses to downhill walking in older and younger individuals. *Journal of Exercise Physiology Online,* 7 (6):45-51.

Keren, G.; Epestin, Y.; Magazanik, A. and Sohar, E. 1981. The energy cost of walking and running with and without a backpack load. *Eur. J. Appl. Physiol.,* (46):317-324.

Kirk, J. and Schneider, D.A., 1992. Physiological and perceptual responses to load-carrying in female participants using internal and external frame backpacks. *Ergonomics,* 35:445-455.

Knapik, J.; Harman, E. and Reynolds, K. 1996. Load carriage using packs: A review of physiological, biomechanical and medical aspects. *Journal of Applied Ergonomics,* 27 (3):207-216.

Knight, C. A. and Caldwell G. E. 2000. Muscular and metabolic costs of uphill backpacking: A hiking poles beneficial? *Med. Sci. Sports Exerc.,* 32 (12):2093-2101.

Laursen, B., Ekner, D., Simonsen, E.B., Voigt, G., and Sjgaard, G., 2000. Kinetics and energetics during uphill and downhill carrying of different weights. *Applied Ergonomics,* 31:159-166.

Legg, S.J. and Mahanty, A. 1985. Comparison of five modes of carrying a load close to the trunk. *Ergonomics,* 28: 1653-1660.

Lin, Y.C., Mao, J.J. and Wang, E.M. 2004. The comparisons of anthropometric characteristics among four peoples in East Asia. *Applied Ergonomics,* 35; 173-178.

Ling, W.; Houston, V.; Tsai, Y.S ; Chui, K., and Kirk, J. 2004.

Women's load carriage performance using modular lightweight load-carrying equipment. *Military Medicine, Nov,* 2004.

Lloyd, R and Cooke C. B. 2000. The oxygen consumption with unloaded walking and load carriage using two different backpack designs. *European Journal of Applied Physiology,* 81 (6):486-492.

Malioy, C.M., Heglund, N.C., Prager, L.M., Cavanga, G.A., and Taylor, C.R. 1986. Energetic cost of carrying loads: Have African women discovered an economic way. *Nature,* 319 (6055) : 668-669.

Mc Ardle, D.E.; Katch, F.I. and Katch, V.L. 1994. Essentials of exercise physiology. Lea and Feblger, A Waverly company Philadelphia. 508 p.

Monod, H and Garcin, M 1996. Use of physiological criteria for improving physical work conditions. *J. Hum. Ergol. (Tokyo),* 25 (1):29-38.

Morgan C.T and Cook, J.C. 1963. Human engineering guide to equipment design. McGraw Hill Book Company Inc., New York. Toronto. London.

Morrison, C.S. and Harrington R.E. 1962. Tractor seating for operator comfort. *Agricultural Engineering,* 43 (11):632-635 and 650-652.

Murrel, K.F.H. 1965. Human performance in industry. Reinhold Publishing Cooperation. New York.

Myles, W.S and Saunders, P.L. 1979. The physiological cost of carrying light and heavy loads. *European Journal of Physiol. Occuo. Physio.,* 42 (2):125-131.

Nag, Anjali.; Nag, P.K. and Desai, Hina. 2003. Hand anthropometry of Indian women. *Indian Journal of Medical Research,* 43 (6): 162-167.

Nigg, B.M. and Herzog W.1999. Bio-mechanics of the muscle-skeletal system. 2nd Edition, John Willey and Sons, New York. 156 p.

NIOSH, 1981. Work practices guide for manual lifting. NIOSH Technical report, Cincinnati.

NIOSH, 1991. Scientific support documentation for the revised

NIOSH lifting equation: Technical contract reports. US Department of Commerce, National Technical Information Service, Springfield.

Panchón, E.; Lobato, R.; Sanchez ,F. and Panchón, A. 2004. Index for quality control in anthropometric surveys. *International Journal of Industrial Ergonomics,* 34 (6): 479-482.

Patton, J. F.; Kaszuba, J.; Mello, R. P. and Reynolds, K. L. 1991. Physiological responses to prolonged treadmill walking with external loads. *European Journal of Applied Physiology,* 63 (2): 89-93.

Patton, J.F.; Kaszuba, J.; Mello, R.P. and Reynolds, K.L. 1990. Physiological and perceptual responses to prolonged treadmill load carriage. United States Army Research Institute of Environmental Medicine Technical Report No. T11-90, Natick, Mass, USA.

Persinger, M.A. 1980: The Weather Matrix and Human Behaviour, New York: Praeger, 327 pp.

Persinger, M.A., 1975. Lag responses in mood reports to changes in the weather matrix. *International Journal of Biometeorology,* 19 : 108-114.

Pheasant, S. 1996. Body space: Anthropometry ergonomics and design of work, 2nd Edn. Taylor and Francis, London.

Pimental, N.A. and Pandolf, K.B. 1979. Energy expenditures while standing or walking slowly uphill or downhill with loads. *Ergonomics,* 22 : 963-973.

Pirnary, F.; Petil, J. M. and Derrone, R. 1996. Compared evaluation of heart rate and body temperature during muscular exercise in hot environment. *Ergonomics,* 28 (2): 28-30.

Pivarnik, J. and Sherman, N. 1990. Responses of aerobically fit men and women to uphill and downhill walking and slow jogging. *Med. Sci. Sports Exerc.,* 22:127-130.

Pund, S.; Yadav, R.; Savani, J.B.; Yadav, S. and Garala, S. 2002. Anthropometry of agricultural workers from Swarashtra region and physiological assessment during spraying operation. Paper presented in XXXVI ISAE Convention, IIT Kharagpur (W.B), India.

Punnet, L. and Herbert, R. 2000. Work related musculoskeletal disorders. Is there a gender differential and if so what does it means? *In*: Goldman, M., Hatch, M. (eds.), Women and Health, pp. 474-492.

Richards, J.H., and C. Marriott, 1974. Effect of relative humidity on the rheologic properties of bronchial mucus. American Review of Respiratory Disease, 109, 484-486.

Robergs, R.; Wanger, D. and Skemp, K. 1997. Oxygen consumption and energy expenditure of level versus downhill running. *J. Sports Med. Phys. Fitness.*, 37:168-174.

Robertson, R.J.; Caspersen, C.J.; Allison, T.G.; Skinar, G.S.; Abbott, R.A. and Metz, K.F. 1982. Differentiated perceptions of exertion and energy cost of young women while carrying loads. *European Journal of Physiol. Occuo. Physiol.*, 49 (1):69-78.

Robinette and Hudson. 2004. Handbook of human factors and Ergonomics, 3rd Edn. Gavriel Salvendy (ed.). Wlley John Wiley & Sons. Inc., Hoboken, New Jersey. pp. 322-339.

Rodahl, K. 1989. The Physiology of Work. Taylor and Francis Ltd., London. pp. 51-79.

Roebuck, J.A.; Kroemer, K.H.E and Thomas, W.G. 1975. Engineering anthropometry method, 2nd Edn .Wiley, New York. pp. 20-46, 173-186.

SAE International. 1998, Executive summary: Civilian American and European Surface Anthropometric Resource–CAESAR (Warrendale, PA: Society of Automotive Engineers Inc.).

Samantha, A.; Datta, S.R.; Roy, B.N.; Chatterjee, A. and Mukherjee, P. K. 1986. Estimation of maximum permissible loads to be carried by Indians of different ages. *Ergonomics,* Dec. pp. 825-831.

Shah, B.; Kathryn S. and Hollenbeck, C.B., 2006. Comparison of Ideal Body Weight Equations and Published Height-Weight Tables with Body Mass Index Tables for Healthy Adults in the United States. *Nutrition in Clinical Practice,* 21 (3): 312-319.

Singh and Kaul. 1972. Human energy requirements of selected farm operations. *J. Agri. Engg.*, 11 (3): 44-52.

Soule, R.G and Goldman, R.F. 1969. Energy cost of load carried on the head, hands or feet. *J. Appi. Physiol.*, 27:687-690.

Soule, R.G.; Pandolf, K.B. and Goldman, R.F. 1978. Energy expenditure of heavy load carriage. *Ergonomics,* 21:373-381.

St. Germain, E.F. 1985. The effects of environmental factors on worker productivity in the construction industry. Mas ter's thesis, Washington University Seattle.

Steele, E.; Bialocerkowski, A. and Grimmer, K. 2003. The postural effects of load carriage on young people–a systematic review. *BMC Musculoskeletal Disorder,* 4 (12):1471-2474.

Stephen, E.; Bechtold, R.E. and Janaro DeWitt L.S. 1984. Maximization of Labor Productivity through Optimal Rest-break Schedules. *Management Science*, 30 (12):1442-1458.

Suggs, C. W and Splinter. 1961. Effect of environment on the allowable work load of man. *Trans ASAE,* 4:48-51.

Taboun, S.M and Dutta, S.P. 1989. Energy cost models for lifting and carrying tasks, *International Journal of Ergonomics,* 4 : 1-17.

Thakur, T.C. and Das, H. 1978. Effects of environments and modes of operation on human performance. *J. Agri. Engg.*, 15 (3) : 151-157.

Tromp, S.W. 1963. Medical Biometeorology, New York: Elsevier.

Wanta, D.; Nagle, F. and Webb, P. 1993. Metabolic responses to graded downhill walking. *Med Sci. Sports Exerc.*, 25:159-162.

Willems, P.A.; Cavangna, G.A. and Heglund, N.C. 1995. External, internal and total work in human locomotion. *J. Exp. Biol.*, 198 : 379-393.

Winsman, F.R. and Goldman, R.F. 1976. Methods for evaluation of load-carriage systems. *Perceptual and Motor Skills.*, 43 : 1211-1218.

Index

